Industrial and Commercial Power System Analysis
Fundamentals and Practice

IEEE Press
445 Hoes Lane
Piscataway, NJ 08854

Industrial and Commercial Power System Analysis Fundamentals and Practice

J. J. Dai
Research Associate/Consultant at University of Tennessee, Knoxville
Irvine, CA, USA

IEEE*Press*
WILEY

Published by John Wiley & Sons, Inc., Hoboken, New Jersey.
Published simultaneously in Canada.

Library of Congress Cataloging-in-Publication Data Applied for:

Hardback ISBN: 9781394265039

Cover Design: Wiley
Cover Image: © Govindanmarudhai/Getty Images

Set in 9.5/12.5pt STIXTwoText by Straive, Chennai, India

Contents

About the Author

J. J. Dai (Senior Member, IEEE) holds BS, MS, and PhD degrees in electrical engineering. He spent 21 years at ETAP, advancing through various roles to eventually become senior vice president. He then joined Eaton Corporation, where he served as director of renewable applications and country manager of engineering services for APAC for 5 years, followed by a 3-year tenure as global account manager for the Oil, Gas & Chemical Group in the United States. His last full-time affiliation was with the U.S. Department of Energy as a physical scientist, focusing on technology management with an emphasis on solar technology integration into the power grid. He is retired and currently works part time as a research associate at the University of Tennessee, Knoxville, where he supports renewable energy projects funded by the National Science Foundation and the U.S. Department of Energy.

J.J. is actively involved in IEEE, contributing to several societies and subcommittees, including IAS, I&CPS, and PCIC. He previously served as the chair of I&CPS Power System Analysis Subcommittee and co-chair of the IEEE 3002.8 Standard Working Group. He currently holds the position of secretary for the IEEE P3002.9 Standard Development Working Group, focusing on power system transient stability analysis in I&C power systems, and chairs the Codes & Standards Committee of IEEE IAS I&CPS. J.J. is also a registered professional engineer in the state of California, USA.

About the Author

Preface and Acknowledgments

Power system analysis is essential for new system design, expansion, and operation. This necessity extends to industrial and commercial (I&C) power systems, which have become increasingly complex in terms of voltage level, size, network configuration, distribution framework, generation sources, and load profiles. The industry has evolved from spreadsheet-based calculations to comprehensive computer software modeling and simulation, with various commercial software packages now available for load flow, short-circuit, protection coordination, motor starting, harmonic, transient stability analyses, and others.

Despite the availability of IEEE standards for recommended practices, a gap exists between industrial standards and college education. Industrial standards do not fully explain detailed modeling and simulation procedures for I&C power system analysis. Meanwhile, college education often lacks sufficient coverage of I&C power systems, with textbooks not adequately addressing fundamentals such as computer modeling, simulation of operating conditions, and study scenario selection.

Having worked for 22 years developing commercial computer software for power system analysis, 8 years leading engineering service teams in the power industry, and actively participating in IEEE standards development for 30 years, I envisioned writing a book to bridge this gap. This book provides essential knowledge from system modeling to computer solutions for I&C power systems, referencing relevant IEEE standards and offering rich analysis examples and illustrations.

The book focuses on:

1) Computer modeling
2) Mathematical solutions
3) Numerical simulations
4) Industrial and commercial power system analysis
5) Industrial standards
6) Computer software applications

It serves as a valuable reference for:

1) College and university students studying power system engineering
2) Engineers engaged in I&C power system analysis
3) College/university teachers
4) Power engineers, certified engineering consultants, and other professionals
5) Trainers and educators in power system engineering

Readers will gain a deeper understanding of system characteristics, modeling and simulation techniques, data requirements and preparation, and modeling tuning and validation. Practical examples from industrial plants and commercial facilities will help readers understand how to build accurate computer models, why simulation studies provide reliable results, and which best practices are recommended by industrial standards. This book aims to prepare college graduates for industry roles and help new or junior engineers become proficient in I&C power system analysis, enabling them to assume greater job responsibilities. By bridging the gap between industrial standards and college education, this book will advance both academic and professional expertise in power system analysis.

This book is organized as follows:

Chapter 1 provides an introduction to I&C power systems. It begins with an overview of single-line diagrams, which form the foundation for system analysis. Key components and equipment in I&C power systems are described, including power supplies from utilities, on-site power generation, power distribution network, and various loads. The chapter also covers power quality conditioning equipment and essential auxiliary systems.

Chapter 2 outlines the objectives and procedures for analyzing I&C power systems at different stages of development and operation. This includes conceptual, preliminary, and detailed design for new and expanding systems, as well as diagnosis and validation for systems in operation. The chapter discusses the required input data, system description data, study scenarios, and the expected analysis outputs for computer modeling and from simulation.

Chapter 3 delves into the detailed computer modeling of I&C power systems necessary for simulation analysis. It examines key components, equipment, control systems, and networks, offering appropriate modeling approaches, such as circuit models, transfer function models, differential equation models, differential and algebra equation models, and Y-Bus and Z-Bus models. The chapter also describes the integration of these models into an overall system model and addresses special operation modes like parallel and isolated operations, along with their associated models.

Chapter 4 focuses on mathematical simulation solutions for major types of I&C power system analyses, including load flow, short-circuit, motor starting, harmonics, and transient stability. Depending on the requirements for accuracy and applicability, some analyses may have multiple solution methods.

Chapter 5 provides fundamental knowledge of numerical techniques used to solve various types of I&C power system analyses. These techniques address algebraic equations, differential equations, integral equations, and mixed linear algebraic and differential equations as formulated in Chapter 4.

Chapter 6 illustrates common and basic I&C power system analysis cases using computer modeling and simulation tools. The cases include analysis and study for voltage drop, power factor correction, circuit breaker overcurrent, reactive power inrush, harmonic distortion, rotor angle stability, and frequency decay and load shedding protection. Some of these cases are based on real-world systems and include description of system configurations, problem formulations, and simulation studies, along with mitigation or protection recommendations.

Appendix A lists the latest IEEE industrial standards for conducting load flow, short-circuit, motor starting, and harmonic analyses and studies of I&C power systems. As the IEEE standard on transient stability analysis is still under development, the appendix includes the latest IEEE paper on this subject, offering a rich and important resource. The analysis scopes, procedures, methodologies, and other key content of each reference are extensively covered.

Finally, I express my sincere gratitude for the understanding and support of my wife, Wen, and my family during the course of working on this book.

August 2024

J. J. Dai
Irvine, California

1

Industrial and Commercial (I&C) Power System

1.1 General Background

Industrial and commercial (I&C) power systems are intricate electrical networks specially designed and operated to deliver reliable electrical power to industrial plants and facilities, as well as for commercial enterprises and buildings. A typical I&C electrical power system comprises several key components or subsystems, including:

1) electric power equipment and controls.
2) lighting systems.
3) power supply for instrument and control systems.
4) air conditioning system.
5) indicating and alarm equipment.
6) receptacles for electrical devices.
7) bonding and grounding systems.
8) emergency, essential, and miscellaneous power systems.
9) Uninterruptible power supply (UPS) systems.
10) Electrical vehicle (EV) chargers.

These systems are composed of various electrical equipment and components, including loads, drives, transformers, lines, cables, reactors, capacitors, switches and protective devices, etc. The systems are powered by either local or regional power utilities or installed on-site generators, or renewable power generations, or a mixture of various energy resources. I&C power systems could be configured in mash or radial, depending on the power supply availability and reliability requirements. Most systems have the capability to reconfigure the network topology to meet different operation scenarios.

The complexity and scale of I&C power systems have grown extensively in recent decades. Some oil and gas fields and facilities could have a total of load amount in gigawatts (GW) level with the voltage rating going up to as high

Industrial and Commercial Power System Analysis Fundamentals and Practice, First Edition. J. J. Dai.

as 220 kV. Three-phase power distribution is commonly employed for high voltage (line-to-line voltage ≥1 kV) systems, while three-phase, two-phase, and single-phase power distributions can be used for lower voltage systems. In normal cases, I&C power systems are connected to power grids, and both are in parallel operation, except in special situations such as on an island, in the desert, or in other remote areas. However, it is not unusual for an I&C power system to temperately separate from the grids and operate in islanding conditions in either planned or unplanned scenarios. For such an operating condition, special design, analysis, and control will be required.

1.2 I&C Power System Single-Line Presentation

For system design, study, and maintenance purposes, I&C power systems are represented by single-line diagrams. A single-line diagram is a schematic that shows the electrical connections of an I&C power system with major equipment and components, including sources, branches, and loads. It is a simplified presentation of a three-phase system due to the symmetry in a balanced three-phase electrical system. Figure 1.1 shows a sample I&C power system single-line diagram that includes a utility connection, two on-site generators, an in-plant distribution system with lines and cables, circuit breakers and switches, transformers, and loads.

1.3 Components and Equipment in I&C Power Systems

Key components and equipment that are needed for successful power supply, distribution, and conversion in I&C power systems are described in the following five categories.

1.3.1 Power Supply and Generation

Power supply and generation units are responsible for electricity to energize and operate local loads. Normally, they include utility or grid supply, on-site generation by conventional generators, and renewable resources and energy storages.

Utility power supply. This is the main source of electrical power supply by local power utility companies through transmission lines or distribution networks. For increasing reliability, multiple lines or even multiple sources can be interconnected to the I&C facility from utilities or grids. Each point of the interconnection is regarded as the point of common coupling (PCC), which usually is on the high-voltage side of the main transformer.

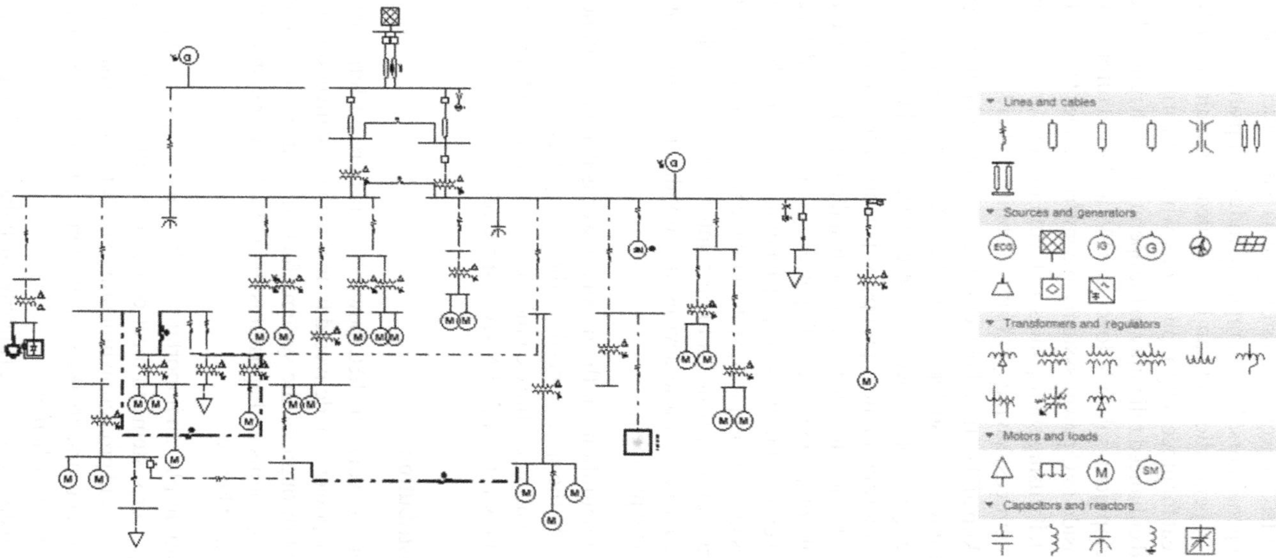

Figure 1.1 A sample I&C power system single-line diagram with symbols for typical components.

On-site power generation. Due to system reliability requirements or economic considerations, large I&C power systems or systems with critical loads often have on-site installed generators. For example, in industrial plants, combined heat and power (CHP) generation set can be installed to simultaneously generate electricity and heat for processing applications, which can bring more efficiency, cost savings, and enhanced reliability to the operation. With renewable resource technology development, solar photovoltaic (PV) and wind turbine systems are also deployed inside I&C power systems to convert energies from sunlight and wind into electricity using solar panels and wind turbines, respectively, which increase the system cost savings and sustainability and create environmental benefits. As backup and optional power sources, diesel and natural gas generation units sometimes are also deployed to the I&C power system.

Energy storage. Energy storage devices such as batteries, flywheels, pumped water, compressed air, thermal or heat pumps, supercapacitors, and others in bio, chemical, or mechanical forms can also be installed and integrated to become a part of critical components at I&C power systems. These devices provide energy and power for power backup, islanding operation, black start, and similar purposes and can improve system efficiency, reliability and resiliency, and green energy support. It needs to be pointed out that the electrical characteristics and controls of these devices are very different from conventional power generation equipment.

The size and configurations of the on-site power generation can be designed and tailored to satisfy the specific needs of the facility. Figure 1.2 shows a typical I&C power system with two utility connections at 46 and 69 kV and two on-site synchronous generators at 5 and 25 MVA, respectively.

1.3.2 Power Distribution

Power distribution networks in I&C power systems are an infrastructure that ensures the electric powers are delivered from the power sources discussed above to loads connected throughout the system efficiently, safely, and reliability. The main electrical components and apparatus in the I&C power system distribution network are described below.

Substation. Substations play an important role in transforming voltage to different levels in I&C power systems to meet equipment ratings and provide various essential operation and protection functions. Substations consist of lines, cables, transformers, switchgears, breakers, protection devices, meters, etc. Figure 1.3 is a sample substation schematic.

Transformers. Transformers are primarily installed to change voltage levels between different clusters of I&C power systems. The utility power imported

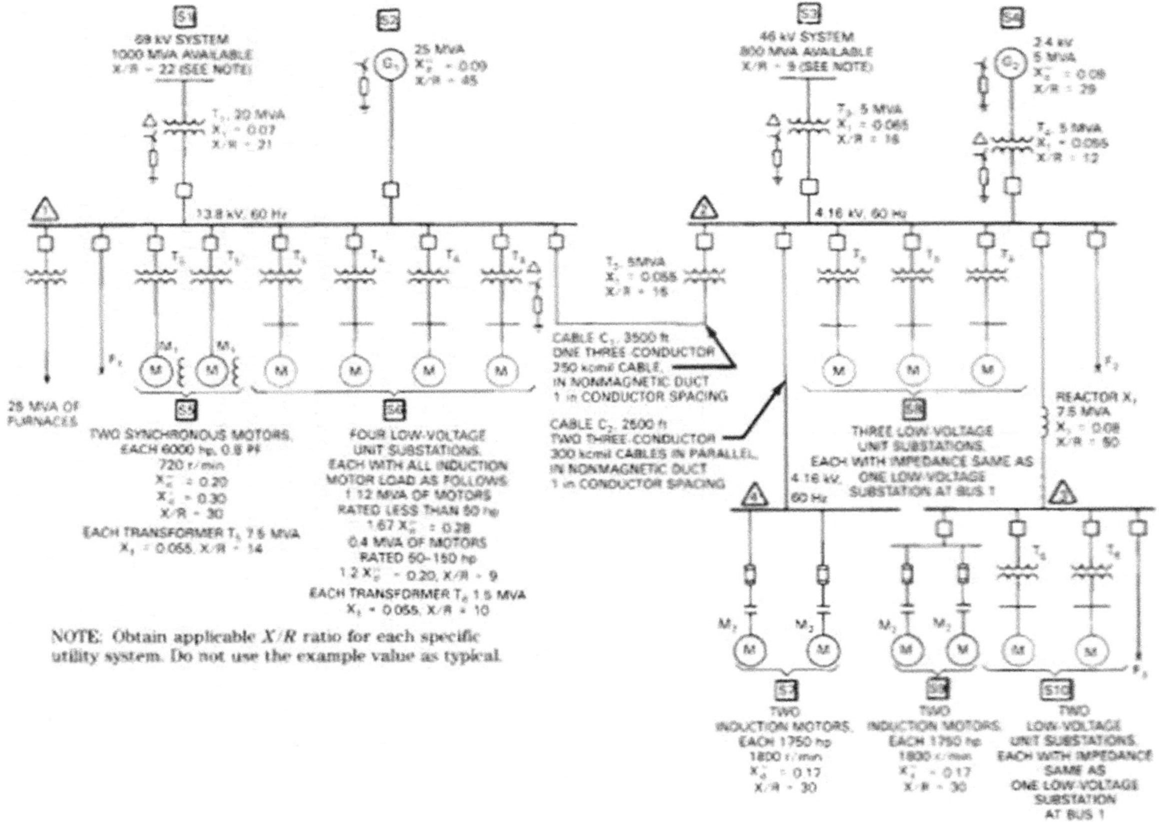

Figure 1.2 Typical I&C power system with utility power supply and on-site generation. Source: Figure 4-10 of [1]. Reprinted with permission from IEEE.

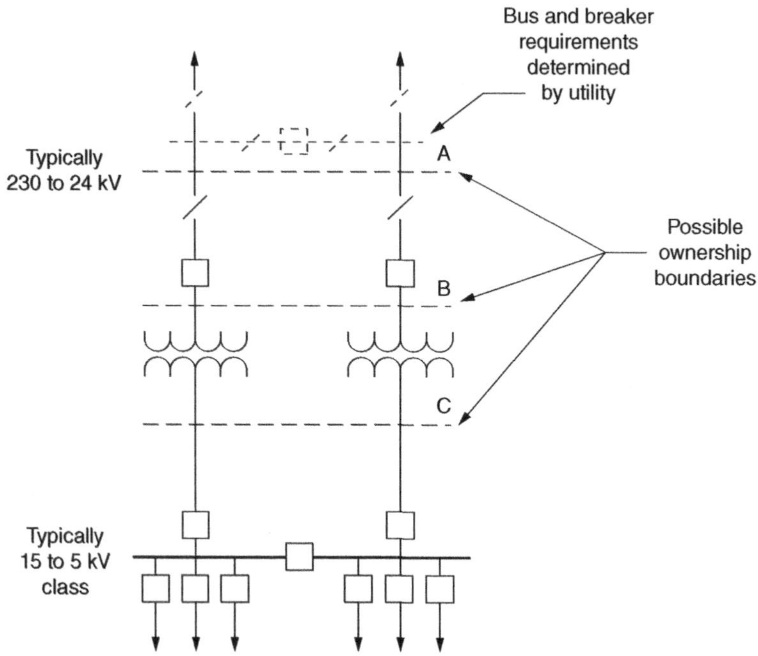

Figure 1.3 A sample substation schematic. Source: Figure 15-2 of [1]. Reprinted with permission from IEEE.

via the transmission system is normally at high voltage and needs to be stepped down inside I&C power systems. Step-up transformers are also used in I&C power systems when on-site generation is at lower voltages. Transformers come with different structures and designs, some are with three-cores, and some are with single core. There are also special-purpose transformers such as ground transformers, zig-zag transformers, transformers with a dedicated phase shift between two sides, open-delta (three-phase windings are in delta connection with open-circuit), etc. From the cooling method point of view, both dry-type and oil-immersed-type transformers are widely used in I&C power systems. Voltage transformer/potential transformer (VT/PT) and current transformer (CT) are also extensively used in I&C power systems, mainly for measuring and protection applications. Figure 1.4 illustrates several typical transformer symbols used in single-line diagrams.

Switchgear. Switchgear is a set of equipment assemblies, including breakers, switches, sensors, inductors and capacitors, and protection devices that are used for controlling and protecting electrical circuits, providing isolation, and safeguarding against overloads and faults. Figure 1.5 is a schematic of a

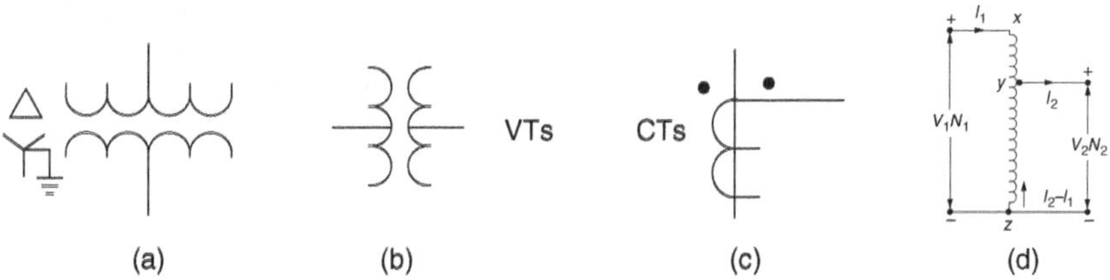

Figure 1.4 Transformer schematic. (a) Power transformer, (b) voltage transformer, (c) current transformer, and (d) auto transformer.Source: Figures 10-1 and 11-3 of [1]. Reprinted with permission from IEEE.

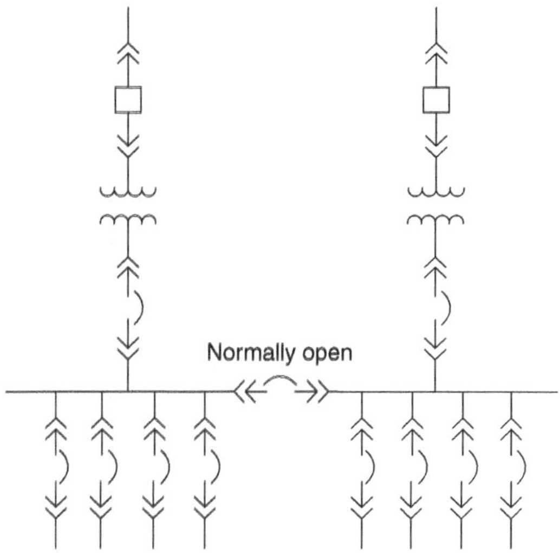

Figure 1.5 A sample switchgear schematic. Source: Figure 2-5 of [1]. Reprinted with permission from IEEE.

switchgear showing only the breakers at both medium voltage (upstream of the transformer) and low voltage (downstream of the transformer) levels.

Distribution lines/cables. Lines and cables form distribution networks in I&C power systems. They connect the power sources to loads through substations, transformers, switchgears and busbars, panels, switchboards. Lines have different structures and installations, including single-circuit or multicircuit; cables have three-phase and single-phase configurations. In general, they are represented as an equipment Pi circuit, as shown in Figure 1.6.

Busbars. A busbar or busduct serves as a common connection point for electrical circuits which form a distribution network within I&C power systems. They are represented as a short bus or a node in a single-line presentation.

Load feeders. Load feeders are directly connected to loads, usually over short distances.

Busbar and load feeder representations are found in Figure 1.7.

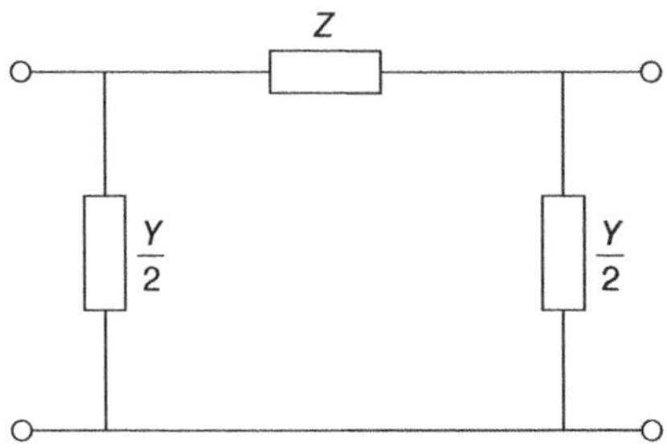

Figure 1.6 Distribution line and cable. Source: Figure 4-8 of [2]. Reprinted with permission from IEEE.

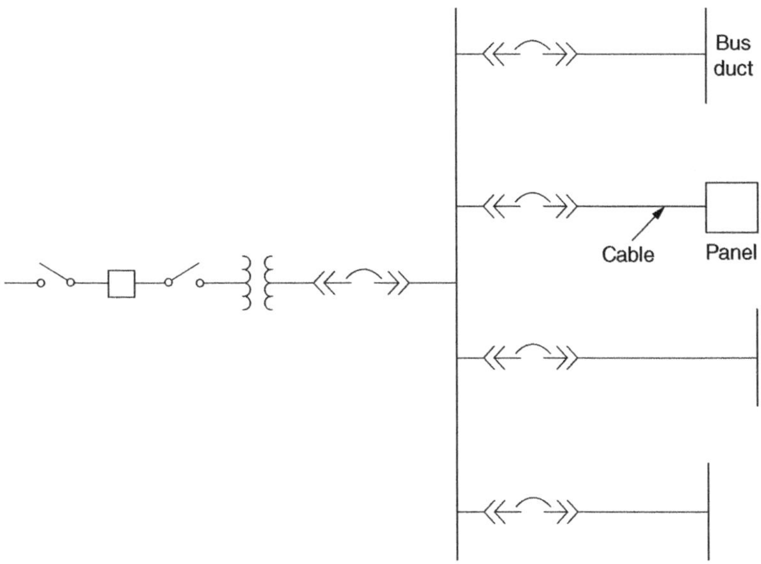

Figure 1.7 Busbar and load feeder. Source: Figure 2-1 of [1]. Reprinted with permission from IEEE.

1.3.3 Loads

Electrical loads in I&C power systems vary widely based on the types of facilities. The primary types of electrical loads are described below.

Motors and drives. Motors make the majority of loads in industrial power systems. They convert electric power into mechanical power to serve numerous processes. There are different types of motors based on the operation principles, including inductor motors, synchronous motors, and DC motors. Motor loads could have very different characteristics, such as a fan, a compressor, a pump, and a conveyor, just to name a few. Modern I&C power systems use motor drives extensively to provide controlled inputs to motors to achieve smoother, more reliable, and efficient operation. Drives are electronic devices with power converters and control circuits.

HVDC. Heating, ventilation, and air conditioning (HVDC) systems are the major loads in both I&C power systems for buildings and special working environments. They employ motors and drives to optimize efficiency.

Lighting. Lighting is found in every I&C power system. There are different designs and technologies for lighting, such as fluorescent lights and light emitting diode (LED) lightings. Their electrical characteristics are also very different.

Other Loads. There are many other electrical loads existing in general I&C power systems with different functions, applications, and characteristics, such as arc furnaces, welding machines, computers, control devices, and communication equipment. Loads can be aggregated into electrical load centers within a specific area and provide a centralized point for power distribution and conversion. Load centers may include electrical panels, subpanels, and branch circuits.

The variety of loads in a general industrial system is illustrated in Figure 1.8.

1.3.4 Power Conditioning

Power conditioning devices are used to mitigate power quality issues and restore sinusoidal waveform for voltage and current. Commonly used devices in this category in I&C power systems include the following described.

Reactors. Reactors are basically used to limit fault currents. They can also be connected in line with the circuits to attenuate harmonic currents or installed inside harmonic filters to work with capacitors to form serious resonant circuits for harmonic filtering. Figure 1.9 shows reactors that are installed in series with generators to limit fault current contribution from the generators into the system.

Capacitors. Capacitors have multiple applications in I&C power systems. The majority of capacitors are installed in shunt with the network to provide reactive power injections into the system to compensate for inductive power consumptions and losses and are called power factor correction capacitors. By the same principle, capacitors provide local reactive power support to reduce voltage drop along the distribution circuits so system voltage profile can be improved. In both power factor correction and voltage drop reduction cases, the capacitor bank can be adjusted in settings either manually or automatically to achieve a regulated control. Capacitors can also be configured with reactors to form harmonic filters with the capacitance turned to specific harmonic frequencies. Figure 1.10 shows an example of various locations to install capacitors with different performance considerations.

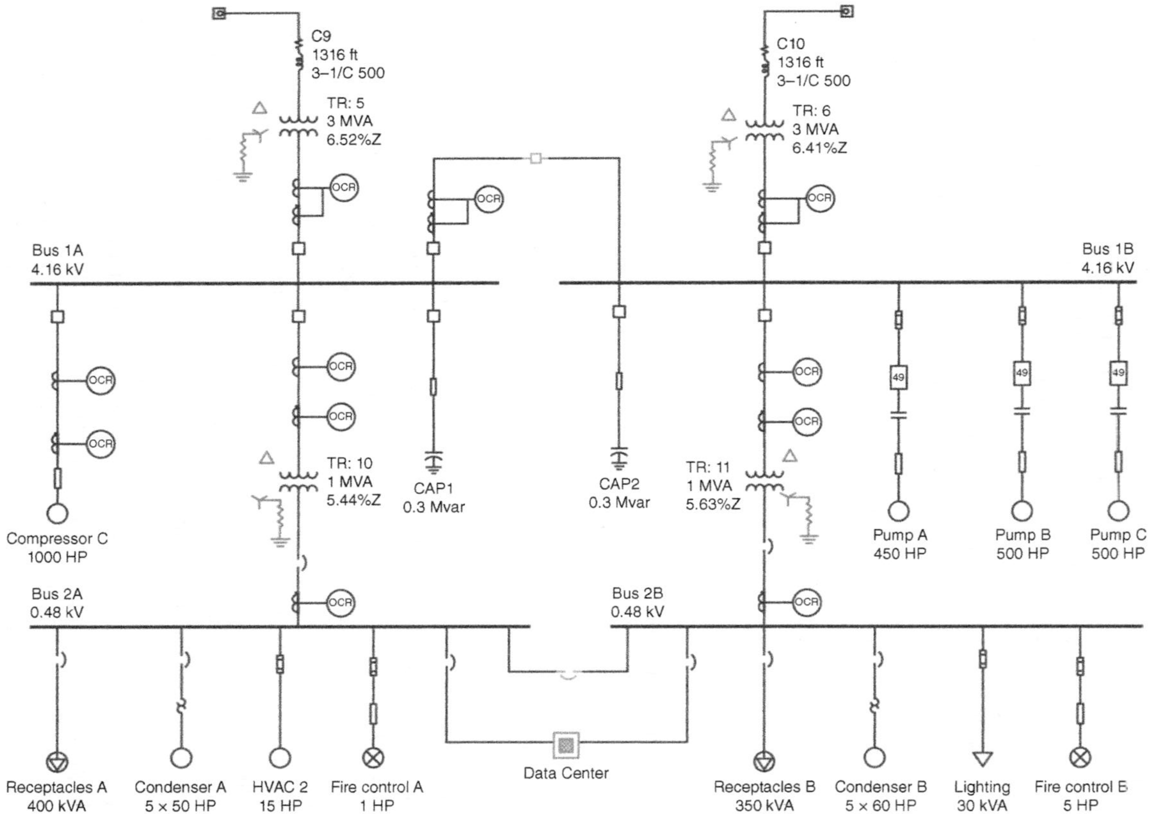

Figure 1.8 Sample substation loads in general industrial system. Source: Figure 6 of [2]. Reprinted with permission from IEEE.

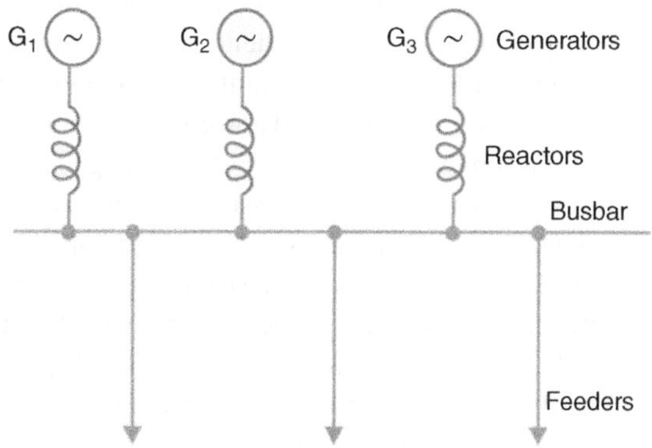

Figure 1.9 Reactor installation example. Source: Types of Current Limiting Reactor, https://www.yourelectricalguide.com/2020/04/current-limiting-reactor.html. Repreinted with permission from yourelectricalguide.com.

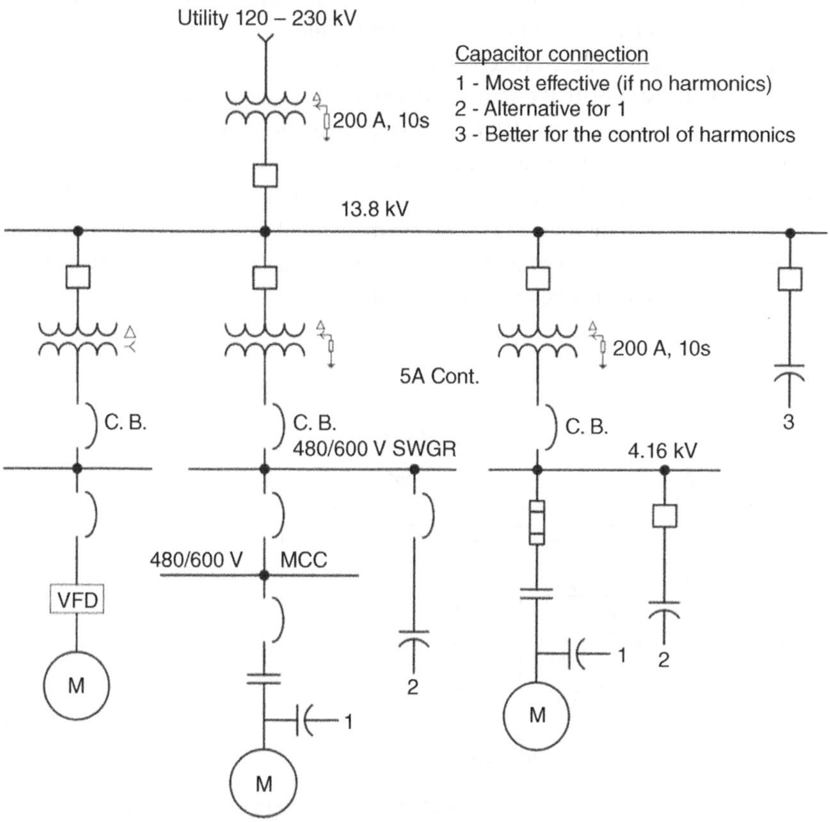

Figure 1.10 Capacitor installation example. Source: How Harmonics Effect Capacitors?!, https://emerichenergy.blogspot.com/2018/09/how-harmonics-effect-capacitors.html. Reprinted with permission from Emerich Energy.

Harmonic filters. Harmonic filters are designed, built, and tuned to filter out currents at designated frequencies, i.e. harmonic currents. By mitigating the presence of harmonics, unwanted and potentially disruptive voltage, and current distortions are reduced and controlled under the limits so the system loads and equipment can operate per the designed objectives. There are two main types of harmonic filters: passive filter and active filter. Passive filters are built using passive components, as introduced above, with reactors (inductors) and capacitors, and sometimes resistors too. Passive harmonic filters have various categories, depending on the tuned filtering frequency points or ranges, such as single-turned, low-pass, band-pass, and high-pass etc. On the other hand, active filters employ power electronic devices to actively inject harmonic currents that cancel out the undesired harmonics. Active filters do not need to be turned to harmonic frequencies and are suitable for harmonic sources with variable frequencies since they automatically adapt harmonic currents for cancellation. Figure 1.11 shows the connection for both the passive filter (and Inductor-Capacitor-Resistor or ICR circuit) and active filter in a hybrid installation.

Uninterruptible power supply (UPS). UPS is a power electronic device with energy storage, typically a battery, plus inverters and control circuits to provide backup power for critical loads in case there is a power outage. UPS is extensively used in I&C power systems and data centers to meet reliability and resilience requirements. Figure 1.12 shows the main elements of a UPS system and the connection configuration between the network and the load.

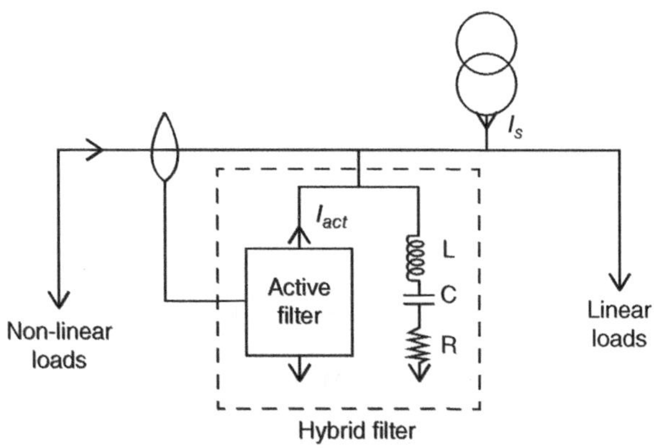

Figure 1.11 Sample harmonic filters installation. Source: https://www.bing.com/images/search?view=detailV2&ccid=uWzNbKC6&id=8925C2D5389F2A5D894773AFF419920B2BDBE33D&thid=OIP.uWzNbKC6uPT-V1ZPAZcUCQHaFy&mediaurl=https%3a%2f%2fuploads-ssl.webflow.com%2f6208452fbd9ac303ffb6bcd1%2f62225b2ac6f789ab96dead88_DHFLPS-07.jpg&exph=1721&expw=2202&q=harmonic+filter+in+industrial+power+system&simid=608009843354178807&FORM=IRPRST&ck=BC5BD5012BB1ED86F5B18EA321FAA61F&selectedIndex=1&itb=0&ajaxhist=0&ajaxserp=0.

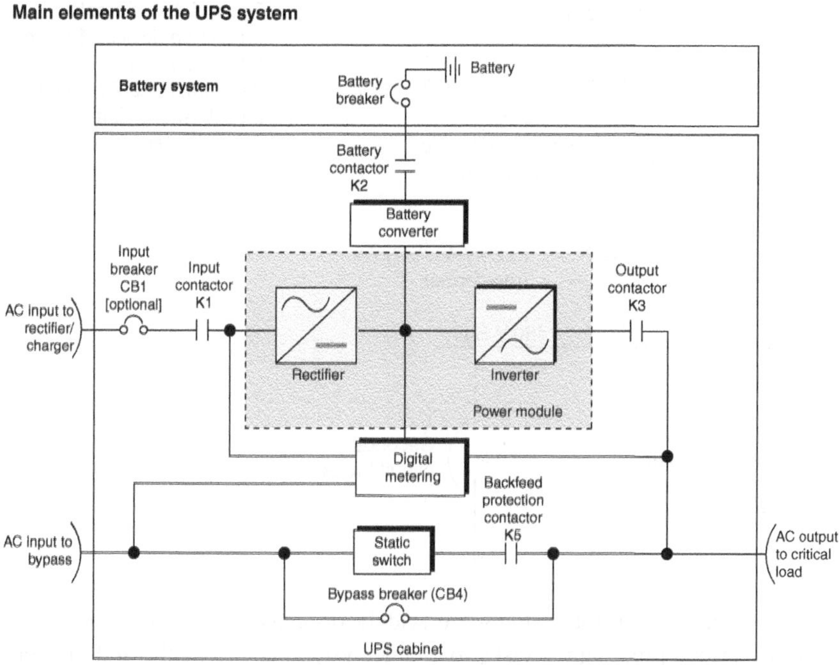

Figure 1.12 Main elements and connection of UPS. Source: Reprinted with permisison from Eaton.

1.3.5 Auxiliary System

Auxiliary systems in I&C power systems are essential to support the safe, reliable, and efficient operations of mail loads. They include some equipment that has been discussed, such as lighting, HVDC, UPS, etc. There are some other equipment and devices that also belong to this category and are described here.

Protections. Protection systems include devices such as current and potential transformers, digital and analog relays, circuit breakers and switches, and fuses to safeguard against electrical faults and overloads, as well as necessary isolation for safety concerns. There are some other devices in I&C power systems that are for different protection functions, including but not limited to surge arrester, snubber, and grounding system. Figure 1.13 depicts several sample devices for the protection of a main transformer and connected loads, including overcurrent relays, a main breaker, a secondary breaker, and a fuse.

Smart meters. Smart or intelligent meters are used to collect data for displaying, analyzing, and monitoring power dispatch and load operation purposes. Some smart meters also have control functions built in. Smart meters are deployed throughout the system at key locations and end users according to the overall system control and monitoring infrastructure.

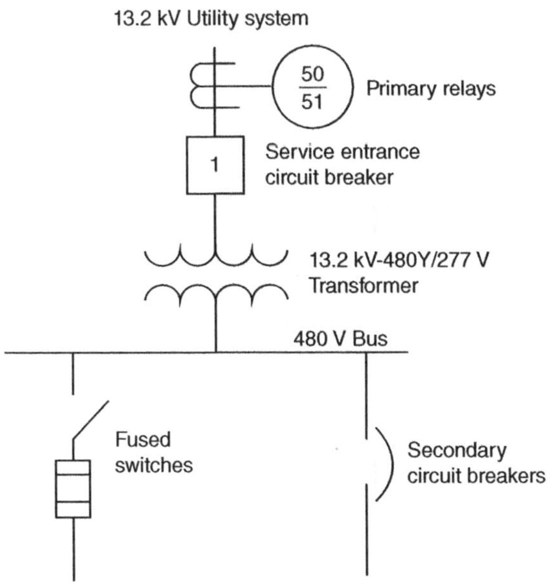

Figure 1.13 Sample protection devices. Source: IEEE Std 399-1997, IEEE Recommended Practice for Industrial and Commercial Power Systems Analysis.

Control systems. Control systems are widely designed and employed to perform various predesigned tasks to enable system safe, reliable, and efficient operations. Typical control systems include programmable logic controller (PLC) for different functions and applications, automatic voltage control (for example, automatic voltage regulator AVR, transformer on-load tap changer [OLTC], etc.), automatic frequency control (for example, generator speed governor, etc.), power factor correction control, motor starting control, arc flash detection and distinction control, adjustable frequency/speed control, load shedding control, distributed control system (DCS), etc.

Monitoring and power management systems. Modern I&C power systems all feature certain data collection and monitoring systems to assess system operating conditions and provide management controls to maintain power quality, stability, and economical operations. Supervisory Control and Data Acquisition (SCADA) systems are widely adapted in power system infrastructure, including I&C systems that provide real-time monitoring and control of industrial processes and commercial operations. SCADA systems are composed of sensors, data transmission networks, real and historical databases, analyzing engines, graphic user interface (GUI) and command dispatching and controls. It will allow system operators and managers to monitor system performance and make decisions. Energy management system (EMS) have some features similar to SCADA in terms of system condition monitoring, data analyzing, decision making, and control command executing with focus on energy-consumption devices and processes to achieve optimal energy efficiency and thus reduce operation costs.

Backup and emergency power generation. Many I&C facilities incorporate backup power systems, such as standby generators, energy storage, UPS, etc., to ensure continuous operation of the system during power outages or provide essential power to critical loads. Some backup and emergency power generation are also responsible for system black start. Direct current (DC) emergency power is also used for circuit breaker control, protective relaying, inverters, instrumentation, emergency lighting, communications, annunciators, fault recorders, and auxiliary motors.

References

1 IEEE Std 141-1993 (R1999), IEEE Recommended Practice for Electrical Power Distribution for Industrial Plants (IEEE Red Book), IEEE, 1999.
2 IEEE Std 399-1997, IEEE Recommended Practice for Industrial and Commercial Power Systems Analysis (IEEE Brown Book), IEEE, 1997.

2

I&C Power System Analysis Applications

2.1 Objectives and Procedures

From planning, to design, and to operations of industrial and commercial (I&C) power systems, system studies are essential steps to assist in the evaluatiing the system performance, reliability, safety and other critical requirements. "Today's intensely competitive business environment forces plant or building management personnel to be very aware of the total owning cost of the power distribution system. Therefore, they demand assurances of maximum return on all capital investments in the power system. The use of digital computers makes it possible to study the performance of proposed and actual systems under many operating conditions. Answers to many questions regarding impact of expansion on the system, short-circuit capacity, stability, load distribution, etc., can be intelligently and economically obtained." [1]. A well trained and experienced electrical power engineer should be responsible for making decisions oversee of which studies are needed for a particular in a specific phase.

Power system study is a full spectrum of complete and detailed engineering procedures to model and analysis the system under various identified real world operation scenarios. Typical and basic studies include evaluation of power supply adequacy to provide energies to loads, system voltage and frequency variations, fault impact and fault interruption capabilities, protections settings and coordination, load/motor start-up and shut-down transients, generator stability margin and protection, power quality issues, energy conversion efficiency, equipment and operation cost, future growth capacity and overall system reliability, etc. These studies are organized into a number of standard recommended power system study and analysis practices and will be discussed in Chapter 4.

Industrial and Commercial Power System Analysis Fundamentals and Practice, First Edition. J. J. Dai.

2.2 Type of Analysis

For typical I&C power systems, system analysis is needed during different system design, operation and upgrade stages. The system can be completely new, existing and in operation, or existing and considered for major expansion. Depending on system status and project phase, the following are categories where system analyses are applied to:

1) Conceptual design
2) Preliminary design
3) Detailed design
4) Multiphase analysis
5) New system
6) System expansion
7) Existing system

2.2.1 Conceptual Design

The objective of a conceptual design analysis for I&C power systems is to search for an optimum power system arrangement in which facility-specific and design-specific criteria are integrated to ensure both present and future operation requirements. In a conceptual study, the emphasis is on the analytical and engineering process of evaluating and verifying the feasibility of several effective alternative designs. During a conceptual design phase for I&C power systems, the basic electrical network infrastructure is to be defined based on the load profiles or specific processes to be supported. The conceptual design phase is often the most crucial in the chain of engineering accomplishments leading up to both new and extensively expanded systems.

System study and analysis are important tools to develop and validate the design concept or improve design performance. The general focus, criteria, tasks and procedures involved in I&C power system conceptual design are described below.

2.2.1.1 Design Focus

The crucial part in these studies is the construction of conceptual design alternatives. This requires broad system engineering expertise, accustomed to recognizing the steady-state and transient performance characteristics of competing designs. This sort of engineering aptitude cannot be developed from textbooks, since there are none covering the subject, and it cannot be developed from working in engineering offices where project engineering production is the primary criterion of performance.

2.2.1.2 Criteria

Following criteria shall be incorporated into the conceptual design within system modeling and analysis area [2]:

1) Power supply adequacy and reliability
2) Power distribution system reliability
3) Equipment reliability
4) Electrical safety
5) Voltage quality and regulation
6) Maintenance requirements
7) Expansion and upgrade flexibility

2.2.1.3 Tasks

The conceptual design of an I&C power system has several tasks [3]:

1) Establish primary and secondary objectives for the project electrical system and classify loads as critical, essential, or general-purpose.
2) Conduct a complete load survey to determine system active and reactive power demands, reliability requirements of the loads from economics point of view, as well as the load types.
3) Specify total connected loads, maximum loads, average loads, and other load categories.
4) Consider and project load growth and system expansion.
5) Evaluate power supply options including co-generation and other on-site generations.

2.2.1.4 Procedures

Conceptual design analysis involves collecting and preparing data, constructing system single-line diagram, developing potential safe, adequate, and cost-effective design concepts to meet the specified design criteria. System simplicity and maintainability shall also be included in the design procedures. Detailed and essential steps involved in the I&C power system conceptual design procedures are addressed below.

1) *Single-line diagram.* Electrical network single-line diagram shall be constructed for building equipment database and conducting analysis. Conceptual design single-line diagram provides a high-level electrical system mainframe, with distribution network structure, power distribution equipment, power supply sources and major or lumped loads. A sample I&C power system conceptual design diagram is illustrated in Figure 2.1.
2) *Load profile analysis.* Loads and associated types with electrical characteristics will be compiled and analyzed for modeling purposes. Connecting points

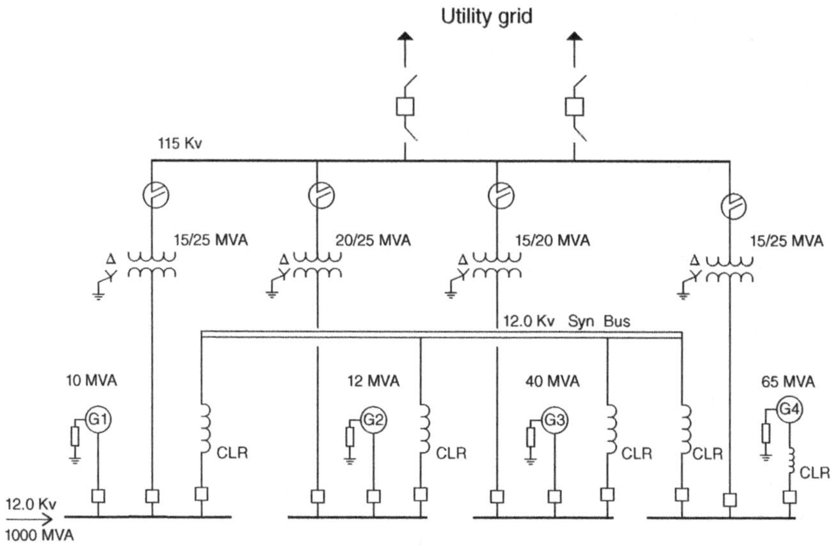

Figure 2.1 A sample I&C power system conceptual design single-line diagram. Source: Dunki-Jacobs and Davis [4]/with permission of IEEE.

and loading categories in different operation procedures will be identified. Future load growth shall also be specified and built into the models. Load data shall include important factors and diversities including connected load, maximum load, average load, future load, etc. Load types and priorities such as motor load, Heating, Ventilation, and Air Conditioning (HAVC) load, lighting load, instrument and control load, auxiliary load, and critical, essential and general-purpose load, etc.

3) *Power source allocation.* Optional power supplies shall be fully investigated, including utility power supply, on-site generation, renewable energy sources and energy storage devices. Advantages and constraints on the availability of each power sources and overall system reliability, resilience, economical operation, efficiency, as well as clean energy commitments will be fully studied and evaluated for each of them with potential mixtures. Special considerations for power source allocation shall take feasibility, efficiency and system stability into consideration, in particular the islanded operating condition, energy availability and cost, anti-disturbance capability and resiliency, and facility constraints, and so on.
4) *Distribution network selection.* Power distribution system configuration and voltage will be selected to best meet system reliability, load centers, facility constraints, and equipment availability. Economic considerations in the distribution network selection shall include power losses, equipment capital cost and operation and maintenance cost, power delivery reliability. System

studies shall assist in comparison and validation of the design selection options to maximize the power supply reliability and efficiency.

5) *Voltage regulation configuration.* Design and validate the system voltage regulation capabilities under various operating conditions by leveraging equipment and technologies such as generator automatic voltage regulator (AVR), transformer on-load tap changers (OLTC) and off-load tap changers, capacitor bank, voltage regulator, static Var compensator, etc.
6) *Harmonic evaluation.* Estimate harmonic loads and harmonic contents. Based on the data, evaluate harmonic penetration into the system and distortion levels inside the system as well as at point of common couple (PCC) with the utility grid. Both harmonic voltage and current distortions at the total level and individual level shall be considered. Identify harmonic sensitive loads, instruments and devices and pay attention to their availability under harmonic distortions. Options to limit harmonic generation sources and harmonic severity shall be explored.
7) *System redundancy and emergency consideration.* Redundancy is a key factor for system reliability, as well as emergency backup power. Different options shall be considered, studied and analyzed to propose optimal solutions which are to be finally validated in the detailed design phase. Cost versus value return shall be a part of the evaluation criteria.
8) *Future expansion plan.* In most cases, system scalability is required. For this purpose, future load growth profiles shall be confirmed to plan suitable expansion margins into conceptual design thus the system infrastructure is expandable to accommodate future loads. In addition to the load growth, generation resource upgrade and replacement, such as renewable generation versus convention generation, shall also be discussed and evaluated.
9) *Other considerations.* The conceptual design shall pay attention to cover other important factors, such as energy distribution and consumption efficiency, safety compliance per standards, control and protection, monitoring and energy management expectations, environment impacts and clean energy commitments, and overall economic incentive and benefits. Some feasibility analysis and studies may be engaged to collect important information.

The conceptual design and study results will serve as the foundation for the next phase in system design, the detailed design, to finish final selection of equipment, installation, protection, and all other technical performance validation.

2.2.2 Preliminary Design

2.2.2.1 Objective

Preliminary design is the next step of the conceptual design. During the preliminary design, the selected conceptual design is to be further studied and detailed with more accurate system and equipment data and broader scope of analysis types for validation and improvement.

2.2.2.2 Focus and Procedures

Scope and requirements for I&C power system preliminary design generally focus on compiling the load profile and performing load flow and short-circuit calculations using preliminary data. Initial equipment selection and sizing from voltage regulation and fault interruption are the main concerns and results from this phase. Steps involved in the I&C power system preliminary design procedures are addressed below.

1) *Single-line diagram.* It shall include all major equipment including utility connections, generations, distribution network, individual or lumped loads, over-current protections, and dynamic components (such as generators and motors) models with initial parameters. A sample single-line diagram for preliminary design is shown in Figure 2.2.
2) *Load profile calculation.* In the preliminary design phase, load data is collected and organized to generate a comprehensive load schedule for individual buses, load centers, and entire system. Equipment nameplate data, vendor datasheet, as well as estimated information are generally used to assemble into load schedules. Both committed loads and spare loads shall be included into the load schedule with proper marks. The loading percentage of each individual or group load shall be defined into practical categories, such as designed, normal, start-up, shut-down, summer, winter, emergency, back up, and so on. Load types are also to be assigned to indicate if the load is constant impedance, constant power, or other types. This information is essential to load flow and other types of analysis. Load schedules may require some calculations, from simple summation to estimation, conversion and interpolation, etc. to cover a full range of loading conditions. For calculations, spreadsheets could be used in some cases, but power system design computer software in general provides special tools to expedite the process with convenience and accuracy. A collection of load schedules forms the system load profile.
3) *Equipment data preparation.* Before performing the other tasks, equipment data required for the preliminary design calculation needs to be prepared. The data can come from the supplier provided datasheet, product categories, qualified computer software libraries, historical data, or estimations. In order to produce accurate calculation results, the equipment data should have come from reliable resources. For the data with certain levels of uncertainties, good tolerances should be considered to reflect variation range of the data so that the calculation results are able to reflect the data tolerances by proving both more conservative and less conservative system evaluations. Note that some equipment data such as rating, and impedance may be adjusted or modified

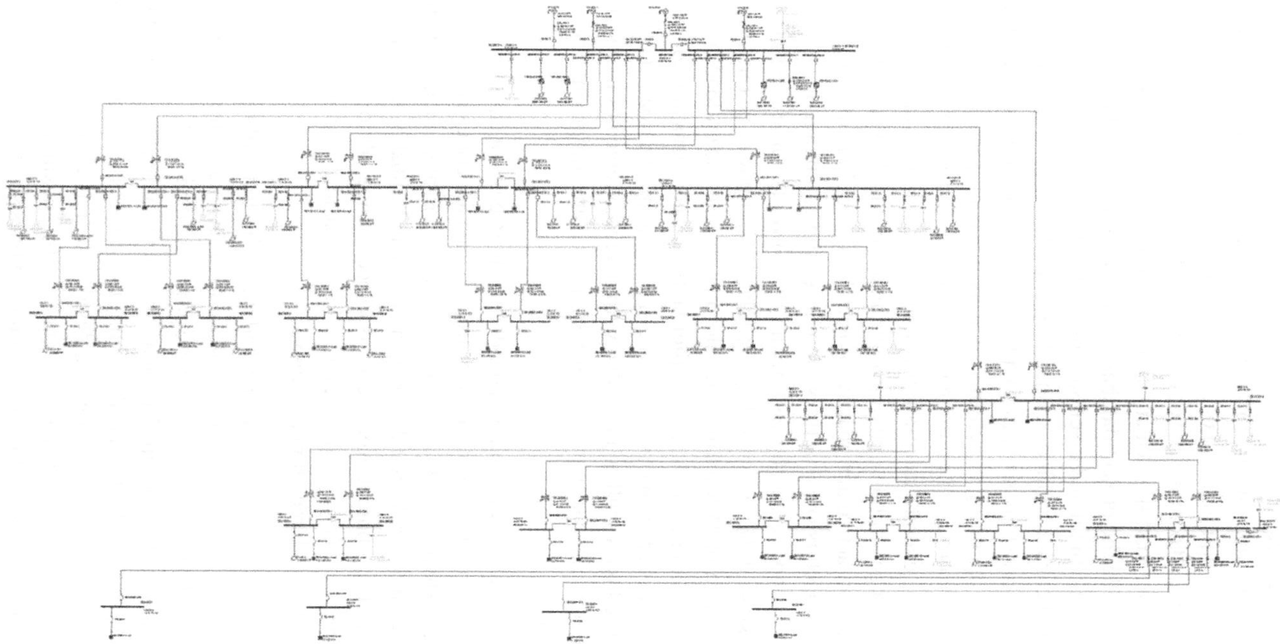

Figure 2.2 A sample I&C power system preliminary design diagram.

from load flow and short-circuit calculations and analysis as system regulation and performance require.

4) *Load flow calculation.* The preliminary load flow calculation is performed after the load profile, equipment data, and network single-line diagram are completed or created. If commercial power system design or other suitable software are used for the load flow calculation, the above data shall be entered into and stored inside the project database. The preliminary load flow calculation should cover the main and essential power systems. The system loading categories should consider different operating conditions, as specified in the load profile. Load flow study scenarios will be designed to assess different normal and abnormal operating conditions such as full load, partial load and overload, equipment outage or maintenance mode, etc. The designed scenarios will be executed to generate calculation results to provide information on power imports, power generations, bus and equipment terminal voltages, branch continuous current, active and reactive power flows, equipment loading, power losses, power factors (PFs) at the system level and equipment level. In particular, voltage drops across the system and the power factor at the PCC will be carefully assessed. Necessary voltage regulation and power factor corrections including generator AVR settings, transformer off-load and on-load taps, capacitor back switching positions, and other static var compensator controls will be evaluated for acceptable performance of the system. Overloaded equipment and excessive losses shall be identified with potential remedies in system configurations, operation scenarios, or by resizing current equipment or installing new equipment.

The preliminary load flow calculation results provide a base for equipment continuous ratings for current, and power. They are also used to derive system reconfiguration schemes to handle credited contingency scenarios, and potential active and reactive load shedding required during generation and imbalance due to generation deficiency. If equipment overloading is to occur under abnormal conditions, corresponding system configuration, generation profile and load profile should be described in detail with recommendations.

5) *Fault current duty and short-circuit calculation.* Short-circuit needs to be calculated during the preliminary design phase in order to determine some important equipment parameters and estimate protective device ratings. Short-circuit calculations are to be performed at each equipment connecting point which has potential exposure to faults, usually a bus or a node in the single-line diagram. All short-circuit current contributing devices such as generators and motors including the spares are considered to generate the maximum fault duties. The system configurations in different operation modes should be considered and those who tend to produce the highest fault duties should be selected as study scenarios. Short-circuit types normally cover three-phase balanced fault and

single-phase ground fault and complete fault duties including momentary duty, interrupting duty and continuous fault duty are to be calculated and evaluated. The evaluation of the fault current duties is used to determine capacity ratings of protective device corresponding to different fault duties. In case of challenges by selecting satisfactory protective device capacities, fault current limiting or reduction solutions shall be considered and evaluated by calculations. Commonly used fault current limitation and reduction devices and solutions include but not limited to current limiting reactors (CLRs), fuses, grounding resistors of transformers, and increasing of transformer and motor impedance. Each of these solutions has pros and cons, for example, applying larger transformer reactance can reduce the fault current magnitude on the path; however, it causes excessive voltage drop and reactive power loss which are not favorable for voltage regulation, power factor and power distribution efficiency.

It is pointed out that during the primary design phase, short-circuit calculation and load flow may need to run in parallel iteratively in order to finalize satisfactory equipment ratings. This is common for other types of I&C power system analysis and during other design and study phases.

6) *Harmonic distortion analysis.* Harmonic distortion shall be analyzed during the preliminary design phase if there are significant harmonic sources existing, e.g., the nonlinear load on a bus exceeding 2% of the short circuit level, or nonlinear load on a bus exceeding 10% of the minimum total bus or feeder operating load, both with respect to the connected bus and in all configurations. All harmonic sources in operation shall be modeled with the harmonic orders up to and including the 65th harmonic. For drives, because their harmonic source spectrum heavily depends on the loading condition; thus, the study scenarios shall cover entire drive's operating speed and power range. The analysis shall calculate total harmonic distortions (THD) and individual harmonic distortions (IHD) for voltage and current components respectively at harmonic sensitive and critical buses, as well as at of PCC with utility and check the distortions against the standard and operation limits. Remedy methods shall be explored, recommended, tested and verified. These methods include but are not limited to harmonic filters, higher pulse drives, etc. More will be discussed in the section 2.2.3. The harmonic distortion calculations shall continue through the project development be finalized before the end of detailed design.

2.2.3 Detailed Design

2.2.3.1 Objectives

The detailed design stage follows the system conceptual design and preliminary design. By this stage, the voltage level, distribution network selection, load profiles, and equipment data have been established. This stage shall continue with the planning of the more detailed aspects of the project to finalize voltage regulation, fault interruption and additional system protection, ensure power quality and stability, and other operation requirements and constraints.

2.2.3.2 Procedures and Analysis Types

I&C power system detailed design needs support from a series of systematical studies to finalize decision on power sources, configuration of power distribution system, selections of equipment rating and size, settings of protection scheme, and other essential system equipment and operation parameters and specifications. Final calculations shall be based on certified vendor data. Following analysis and study tasks are specifically related to the above objectives during the detailed design phase.

1) *Single-line diagram.* A final electrical system single-line diagram includes power generation and distribution network and equipment, control and protection system and devices, configuration and connection of all the above equipment and devices. Figure 2.3 is a detailed design single-line diagram for a sample I&C power system.
2) *Final and detailed load list.* The final load data shall be certified and checked for completeness. All data shall come from credited sources such as equipment nameplates, vendors' and suppliers' datasheets, commissioning tests. Ratings of each load shall provide information to cover a full range of operating conditions from 0% to 100% or higher. The information needed includes but is not limited to current, power factor, efficiency, active power and reactive power. Load service factor and average running time (specified as continuous, intermittent or stand-by for example) are also essential data to confirm. For motor loads, motor circuit model or equivalent data, inertia and load torque data shall also be provided for motor starting and transient analyses. The final and detailed load list is essential for all analysis in the detailed design phase.
3) *Equipment list.* The final equipment and data shall be confirmed. Equipment ratings and parameters shall come from nameplates, vendors' or suppliers' datasheets, commissioning tests, or other validated sources. Control systems or adjustable devices associated with the equipment shall be fully available for both steady state and dynamic state operations. Examples of those are generator governor and excitation/AVR systems, transformer on-load tap changers (OLTC) and off-load tap changers, Var compensators, capacitor switching, and motor starting devices. It is particularly important that models describing control and dynamic behaviors of those components shall be legitimately verified and validated. Parameter tolerance and variations under certain conditions are allowed since many computer simulation software packages can

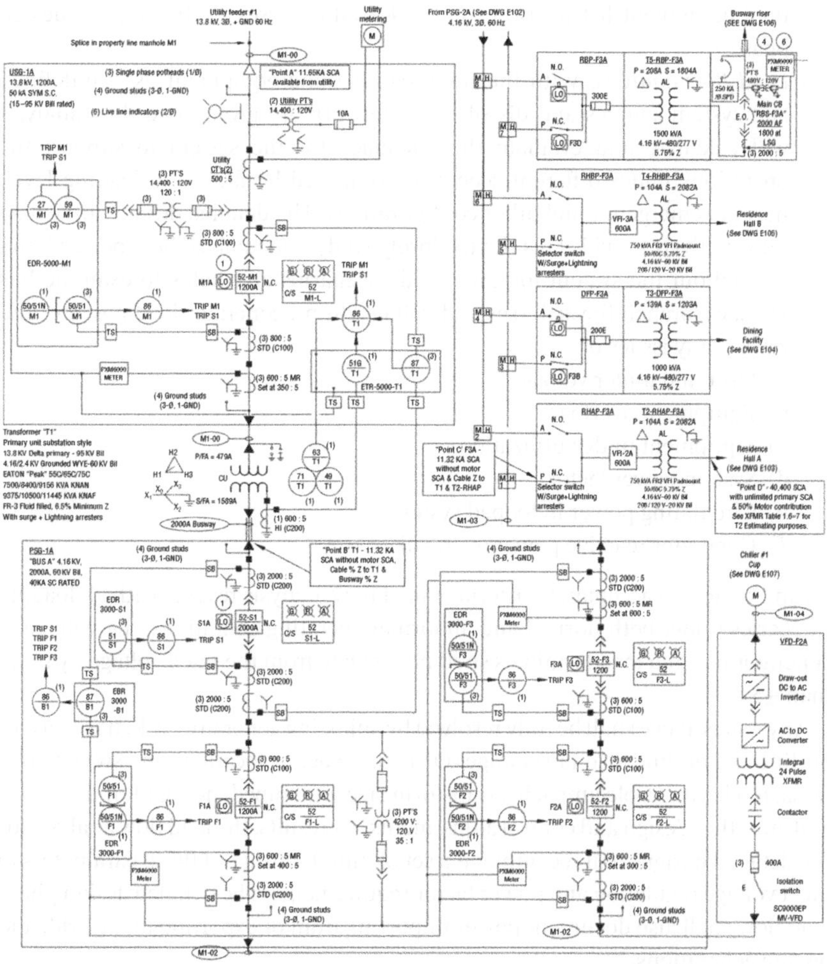

Figure 2.3 A sample I&C power system detailed design diagram. Source: https://www.bing.com/images/search?view=detailV2&ccid=BELQRDXV&id=67F8AABCBF737651217887AF8361C24C47B5FF20&thid=OIP.BELQRDXVvAgENqq_GUkiiQHalg&mediaurl=https%3A%2F%2Felectrical-engineering-portal.com%2Fwp-content%2Fuploads%2F2019%2F07%2Fsingle-line-diagram-power-system-1.gif&cdnurl=https%3A%2F%2Fth.bing.com%2Fth%2Fid%2FR.0442d04435d5bc080436aabf19492289%3Frik%3DIP%252b1R0zCYYOvhw%26pid%3DImgRaw%26r%3D0&exph=1669&expw=1454&q=single+line+diagram+power+system+detailed+design&simid=607988712077814547&form=IRPRST&ck=33288C839BB83C700A2F0C01552D1731&selectedindex=56&itb=0&ajaxhist=0&ajaxserp=0&vt=0&sim=11/with permission of Chegg Inc./Last accessed on Sep 05,2024.

automatically or manually handle the equipment parameter tolerances. The final equipment list is also essential for all analyses in the detailed design phase.

4) *Load flow analysis.* With the final load list and equipment list are defined and system sources and distribution network are given, load flow analysis will calculate and evaluate the adequacy for the system to support the operations under different scenarios composed by factors of loading level, normal, emergency and other configurations. The detailed design phase load flow analysis provides results to closely study and analyze how power flows throughout the system for each study scenarios and helps to examine bus voltages, branch flows, losses and voltage drops, generator loadings, etc. and reveals any issues of:
   a) Over and under voltage
   b) Equipment overload
   c) Equipment under-utilization
   d) System power factor
   e) Circulating power in looped systems
   f) Generator reactive power overload

As an example of acceptance criteria of load flow analysis results, with all loading categories under both normal and abnormal operating conditions, the voltage at generator bus and load terminals shall not deviate from the rated voltage by more than ±5%.

Any issues above shall be fully resolved by adjusting generator AVR, transformer off-line and on-line load tap change (OLTO), capacitor locations and switch positions, modifying cable ampacity, transform size and impedance, CLR inductance, and any other design and control adjustments. As results, the load flow analysis for the detailed design shall recommend a set of initial settings of these parameters to meet voltage regulation and power factor requirements. The detailed design phase load flow shall also document power importing from or exporting to the grid, and on-site generations.

A sample single-line load flow study result is shown in Figure 2.4 in which system solved bus voltage, line power flow are displayed with generation and load connection. The highlighted table indicates the overloaded transformer equipment ID and loading percentage (145% of the rated loading).

5) *Steady-state contingency/security and reliability analyses.* Steady state contingency analysis specifically investigates predefined or auto-defined equipment outage cases to perform a post-contingency load flow study and compares the result to the base condition, i.e., the normal or pre-contingency load flow result. It then generates a rank for each case according to the severity level of the contingency. The information is for engineers to select cases

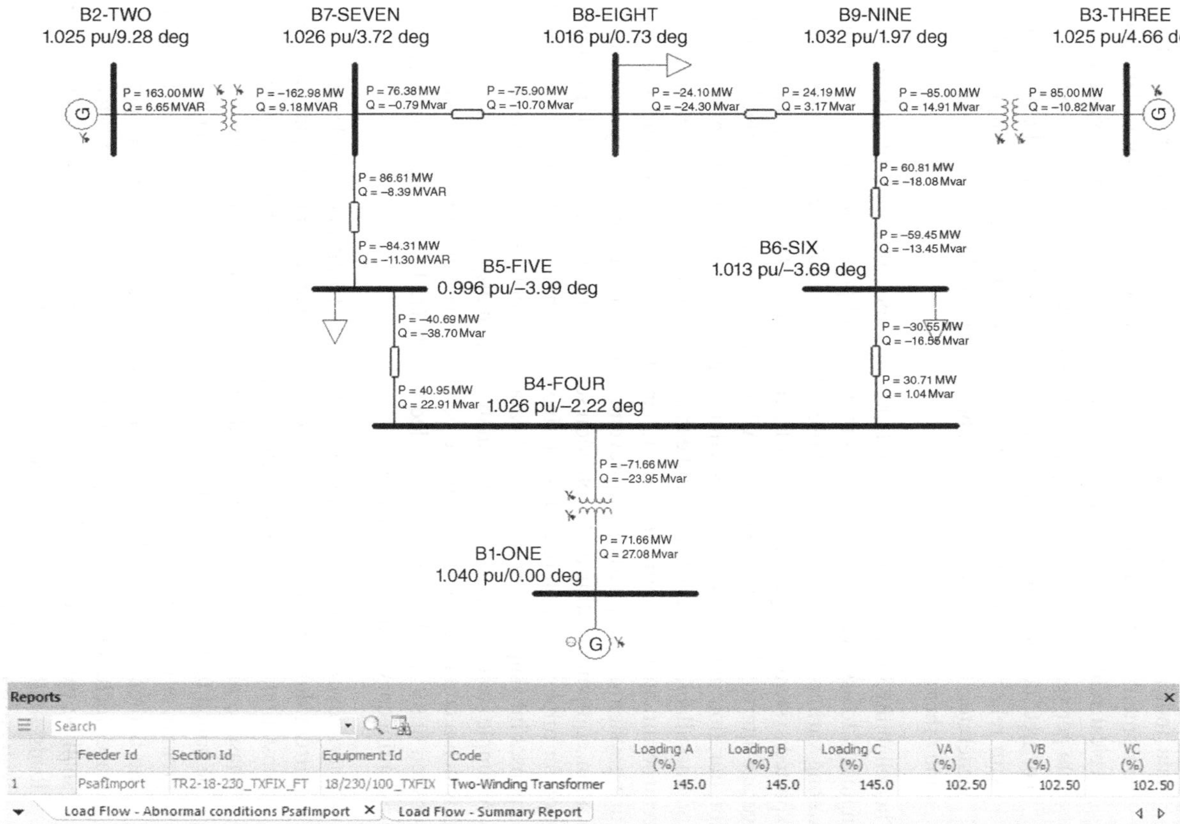

Figure 2.4 Sample load flow result display.

and implement needed remedy actions. Instead of analyzing pre-defined outages of components in the system, reliability analysis calculates system reliability indices based on the reliability data of each component. Reliability data indicates the statistic failure, repair and replacement probability of equipment from published standards, vender's or supplier's datasheet. Typical reliability data include active failure rate per year, passive failure rate per year, mean time to repair time, switch time, and replacement time. For I&C power systems, practically all power equipment can be associated with reliability data and affect the reliability indices calculation, including generator, transformer, line and cable, motor and load, switch and breaker, and others.

The reliability indices provide detailed and, in some cases, quantified values for engineers to assess probability of loss at each load point and prepare solutions. Among various reliability indices, those are often used in I&C power systems, expected system average interruption frequency index (SAIFI), expected system average interruption duration index (SAIDI), expected customer average interruption duration index (CAIDI), expected average service availability index (ASAI) and unavailability index (ASUI), and expected interruption cost (ECOST), etc. As it can be anticipated, contingency and reliability analyses are of great help not only during system design phase for improvement and enhancement in system configuration, equipment rating and protection schemes, but also at the real-time operation for implementation of corresponding system handling by pre-programmed responses into supervisory control and data acquisition (SCADA) systems.

A sample steady-state contingency analysis result display is shown in Figure 2.5 in which a contingency of line outage between Bus 4 and Bus 5 causes the line between Bus 4 and Bus 6 to be overloaded to 129% (right figure) comparing to the pre-contingency loading (left figure).

6) *Short-circuit and fault analysis*. Short-circuit and fault analysis calculates fault current duties for various fault types at each equipment terminal, bus, and substation. Fault types shall include three-phase fault, phase-to-phase fault, phase-to-phase-to-ground fault, and single-phase-to-ground fault. Faults can be bolted fault, meaning no impedance between the faulted circuits and the ground, or impedance connected fault where there is an impedance between the faulted circuits and the ground. All fault duties at critical points and stages including momentary (peak), interrupting (breaking) and steady state shall be calculated, and all operating loads shall be considered. Operating loads shall be configured to generate both the maximum and minimum fault currents. Calculations shall follow applicable standards in treating the machine impedance, selecting the pre-fault voltage, and adjusting the fault current magnitudes based on the nature of the sources.

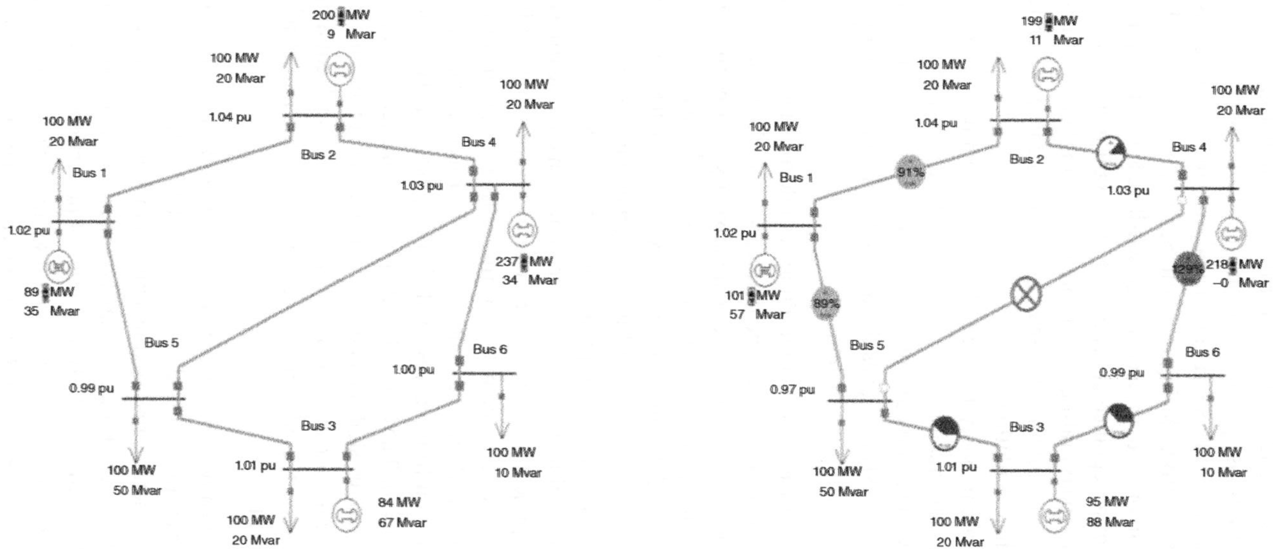

Figure 2.5 Sample steady-state contingency analyses result display. Source: Mishra and Khardenvis [5]/with permission of IEEE.

Calculated fault current duties will be used to select appropriate protective devices such as circuit breakers and fuses to interrupt the fault currents and isolate the faults to ensure continuous system operations. The fault currents are also used in protection coordination study. Short-circuit analysis will also provide useful information of bus voltage drops during the fault. Such information can be used for voltage drop study in order to protect voltage sensitive loads and equipment.

As an example of acceptance criteria of short-circuit analysis results, circuit breaker momentary (peak) rating, interruption (breaking) rating, and dc current rating shall typically maintain some margin (say 10%) above the corresponding fault duties from the calculation.

A sample short-circuit analysis result display is shown in Figure 2.6 in which each bus is faulted individually at a time. The Reports window shows summary of the low voltage (LV) bus fault currents, and the Short-Circuit American National Standards Institute (ANSI) Box gives the detailed fault currents in different fault types at various important time instants.

7) *Protection and device coordination.* Protection devices pick up fault current through relays, current sensors and detectors as for circuit breakers or melting point as for fuses. Some of these are auxiliary devices that are built-in, and some are external. Once fault currents are calculated, protection curves, which are also called the time-current-curve (TCC), shall be created to decide settings of the circuit barkers and fuses together with the auxiliary devices. The principle is to ensure the protection curve intersects the fault current and provides a tripping time less than the minimum fault clearing time determined per the project or corporate procedures. Overloading conditions shall be protected in the same manner. For this purpose, equipment characteristic curves including thermal damage curves in terms of current square versus time and withstand-limit curves shall also be plotted on TCC such that the protective curves shall be below the equipment characteristic curves. Finally, the protection curves shall be above motor starting curves and transformer inrush current curves to allow start up and energize of such equipment. Time difference between the adjacent curves shall meet the minimum clearance time requirement.

Protection curves must be coordinated in their settings to achieve desired operation sequences. For example, it is always desired that the protective device closest to the fault location operates first so that a minimum portion of the operating system will be interrupted. The protection can be done in zones, but the entire system shall be protected. Figure 2.7 illustrates zones of protection in a single-line diagram. Completing this process generates a full setting table for all main protection and auxiliary devices. For some unachievable coordinations and zones, recommendation of mitigation shall be provided.

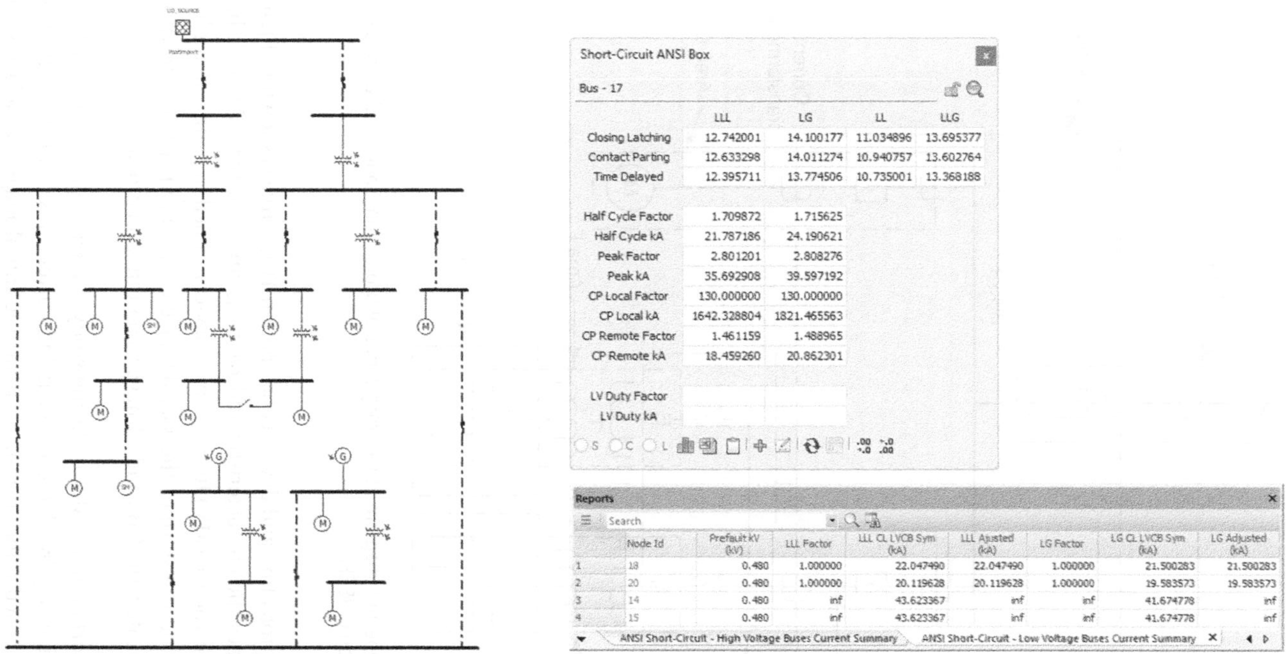

Figure 2.6 Sample short-circuit and fault analysis result display.

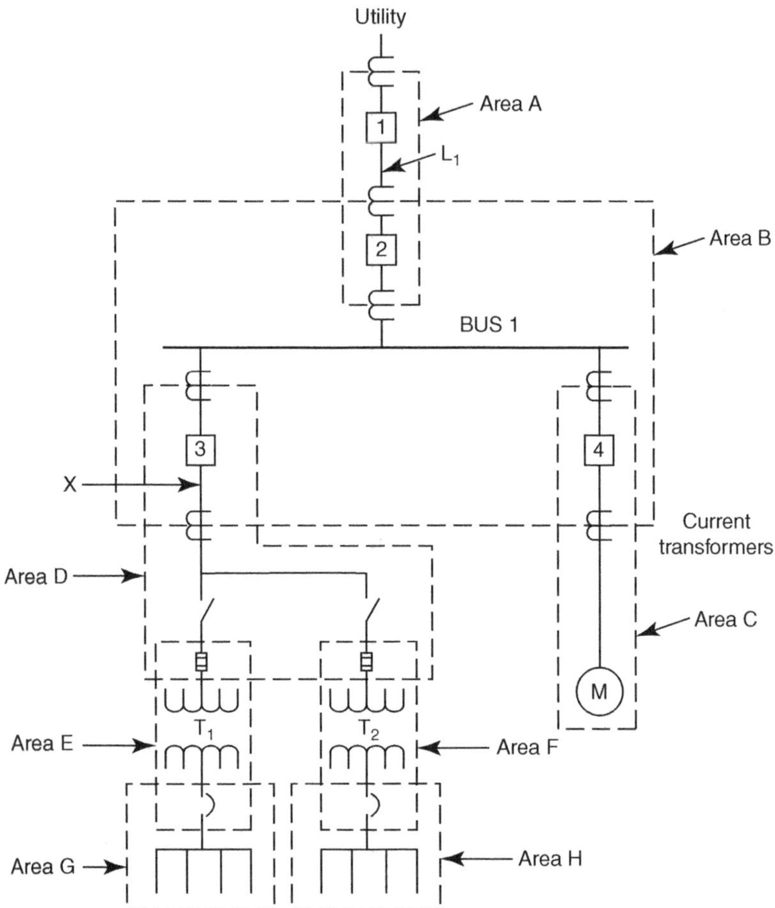

Figure 2.7 One-line diagram illustrating zones of protection. Source: IEEE Std 141 [2]/ with permission of IEEE.

A sample protection and device coordination TCC curve is shown in Figure 2.8 which includes the characteristic and tripping curves of the LV circuit breaker CB1, medium voltage (MV) fuse F1 and relay-controlled medium circuit breaker CB2. Cable and transformer thermal damage curves are partially shown in the TCC. As indicated, the CB2 and F1 have a coordination problem.

8) *Motor start analysis*. Motor staring study simulates motor energization process which requires large magnetizing current, also called the inrush current or locked-rotor current, to build the initial magnetic field insider the motor air gap in order to convert electric energy from the stator winding to the mechanical shaft via interactions with the rotor winding. Due to the network

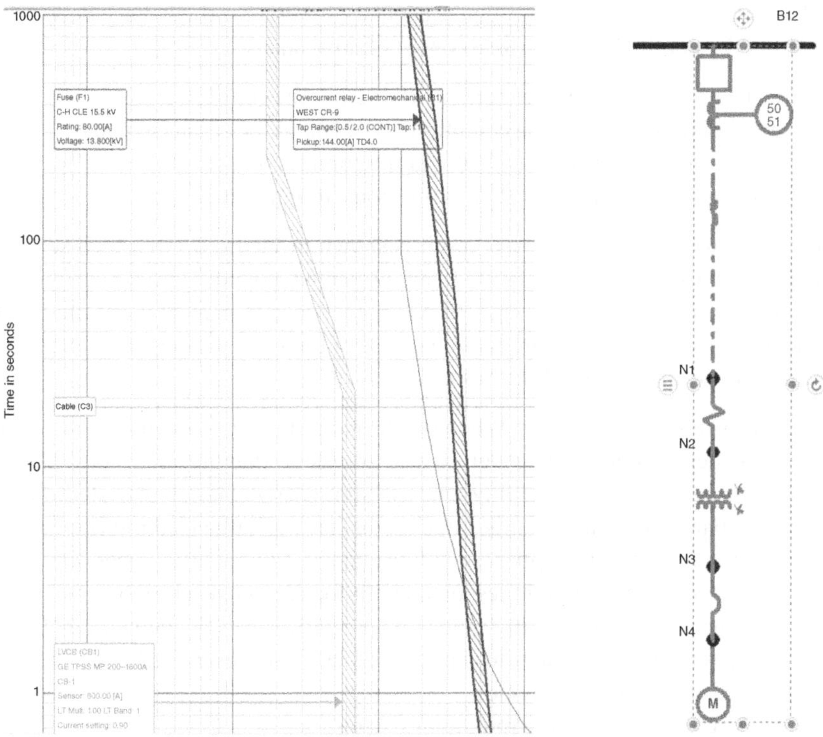

Figure 2.8 Sample protection zone and device coordination result display.

impedance from the source to the starting motor, voltage drops will occur across the network with efferent levels of severity depending on the electrical distance away from the starting motor. I&C power system motor starting analysis has two objectives, the first is to confirm successful accelerations of the starting motors, and the second is to access voltage drops at different locations and their impacts to various equipment in operation. The detailed motor starting analysis shall include the worst scenarios in starting reactive power and inductive current demanding, usually starting the largest motors and in simultaneous or sequential motor starting, as well as any starting procedures per requirements from the process. Voltage drops at generator terminal bus, substations, critical or voltage sensitive load points, and the starting motor shall be carefully assessed and compared to the standard recommended values or corporate design guidelines.

System design engineers shall use the information from the motor start analysis in the detailed design phase to recommend and validate mitigation such as starting device and reduced starting load to improve the voltage profile. The final design shall satisfy voltage drop, locked-rotor current and starting time constraints.

For I&C power systems, common acceptance criteria for the voltage deviation dV from the nominal voltage value consist of three states:

a) During normal operating condition: $dV < 5\%$ at the connected bus.
b) During abnormal operating condition: $dV < 10\%$ at the connected bus.
c) During starting transient, $dV < 30\%$ at the connected bus.
d) During motor staring, $dV < 20\%$ at the motor terminal bus.

A sample motor starting analysis result is shown in Figure 2.9 in which several starting curves of a motor ID 10307-M are presented, including motor current on the base of the full load amperage (FLA), motor speed in percentage of the rated synchronous speed, motor torque on the base of the rated torque, and motor terminal voltage on the base of the bus nominal voltage.

9) *Transient stability analysis.* When synchronous machines are installed and/or for systems of in-plant generation, a transient stability analysis shall be performed. Transient stability analysis simulates scenarios of system large disturbances including faults, large motor starting, large load impact, loss of generation or utility supply (islanding), major equipment outages, etc. and evaluates system dynamic responses with focuses on synchronous machine rotor angle, system frequency and voltage profile across the network. Following system and equipment responses shall be closely examined to determine if there are any stability issues:
   a) Synchronous machine rotor angle swing
   b) Exciter and governor responses
   c) Nadir and rate of change of frequency (RoCoF)
   d) Maxima and minima of voltage variation
   e) Motor speed and torque changes

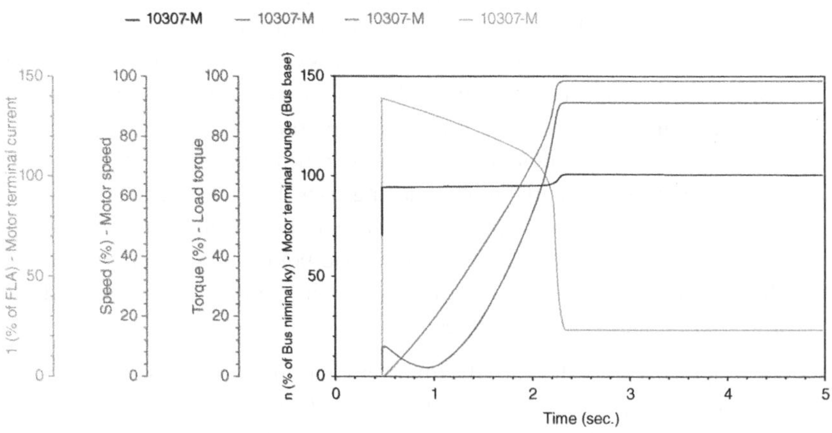

Figure 2.9 Sample motor start analysis result display.

f) Current and power oscillations, and settle-down time of these variables
g) Reactions of voltage and frequency dependent loads such as motors that are running
h) Status changes of protection devices, in particular circuit breakers, voltage and frequency relays

Additionally, the transient stability analysis shall determine the following stability constraints:

a) System total load as well as critical load limitations for each study scenario to maintain system stability
b) Load shedding necessity and configuration in system islanding case to maintain system stability
c) Maximum motor load re-acceleration capability to maintain system stability and voltage regulation

If computer software is engaged, simulation time step shall be set to offer sufficient resolution for the numerical solution algorithm in order to truly reflect machine and other control system dynamics. The total simulation time shall be chosen according to the study scenario for sufficient length of time to determine whether instability exists.

The purpose of transient stability analysis is to ensure studied I&C power system can handle disturbances without collapsing or significantly impacting the operating processes and loads. Special attention shall be given to faulty conditions on the main buses and generator terminal buses, and the action and timing of bus tie breakers operation for stability and protection. Engineers shall determine for each internal fault scenario the critical fault clearing time (CFCT) which is defined as the time to clear the fault as well as for each utility side fault scenario the critical system separation time (CSST), both are constrained by the stability requirement of without rotor angle slipping pole for any synchronous machine in the interconnected system. Stability margins for each scenario and the overall system shall be recommended and established by adjusting system operation levels, power transfer impedance and equipment parameters including those in their control systems. Protection devices and settings including breakers, differential relays, out-of-step relays, and other relays to detect instability and protect illegal operations shall also be configured or recommended if necessary. Requirement and amount of active and reactive load shedding to respond islanding condition shall be recommended if necessary to maintain system stability.

IEC 60034–1[6] defines the voltage and frequency range acceptance criteria during the transient as illustrated in Figure 2.10 and stated in following: "A machine shall be capable of performing its primary function within zone B, but may exhibit greater deviations from its performance at rated voltage and frequency than in zone A. Temperature rises may be higher than at rated voltage and frequency and

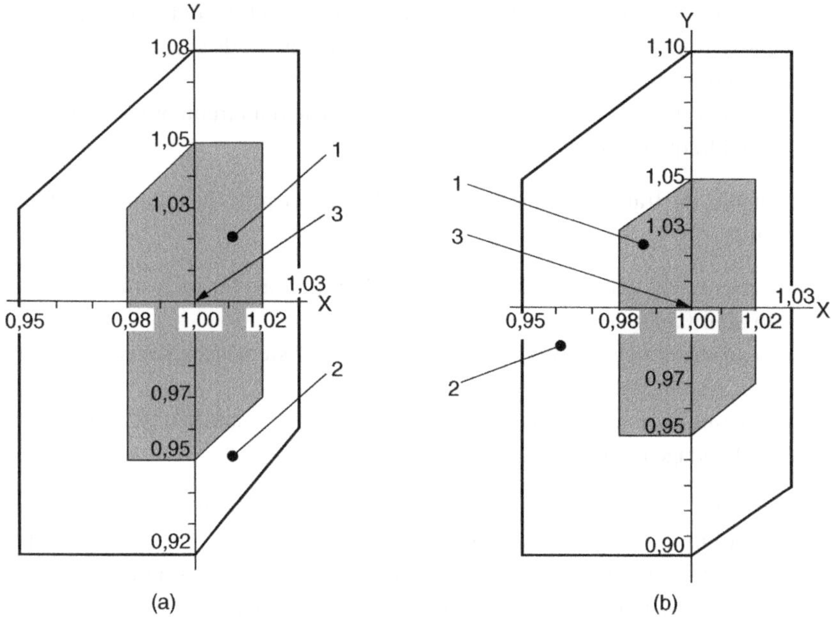

Figure 2.10 Voltage and frequency limits for generators (a) and for motors (b). (X axis: frequency p.u., Y axis: voltage p.u., 1: zone A, 2: zone B [outside zone A], 3: rating point).

most likely will be higher than those in zone A. Extended operation at the perimeter of zone B is not recommended."

Commonly accepted criteria for the voltage and frequency deviations dV and df from the nominal values are given separately in the following for three different operating conditions:

a) During steady state normal operating condition

$$dV < 5\%$$
$$df < 2\%$$

b) During steady state abnormal operating condition

$$90\% < V < 95\% \text{ and } 105\% < V < 110\%$$
$$95\% < f < 98\% \text{ and } 102\% < f < 103\%$$

c) During transient condition

$$70\% < V < 90\%$$
$$92\% < f < 95\% \text{ and } 103\% < f < 110\%$$

All values are based on the equipment ratings.

A sample transient stability analysis result is shown in Figure 2.11 in which a sequence of events is simulated as given in the list. Rotor angle swing of the three

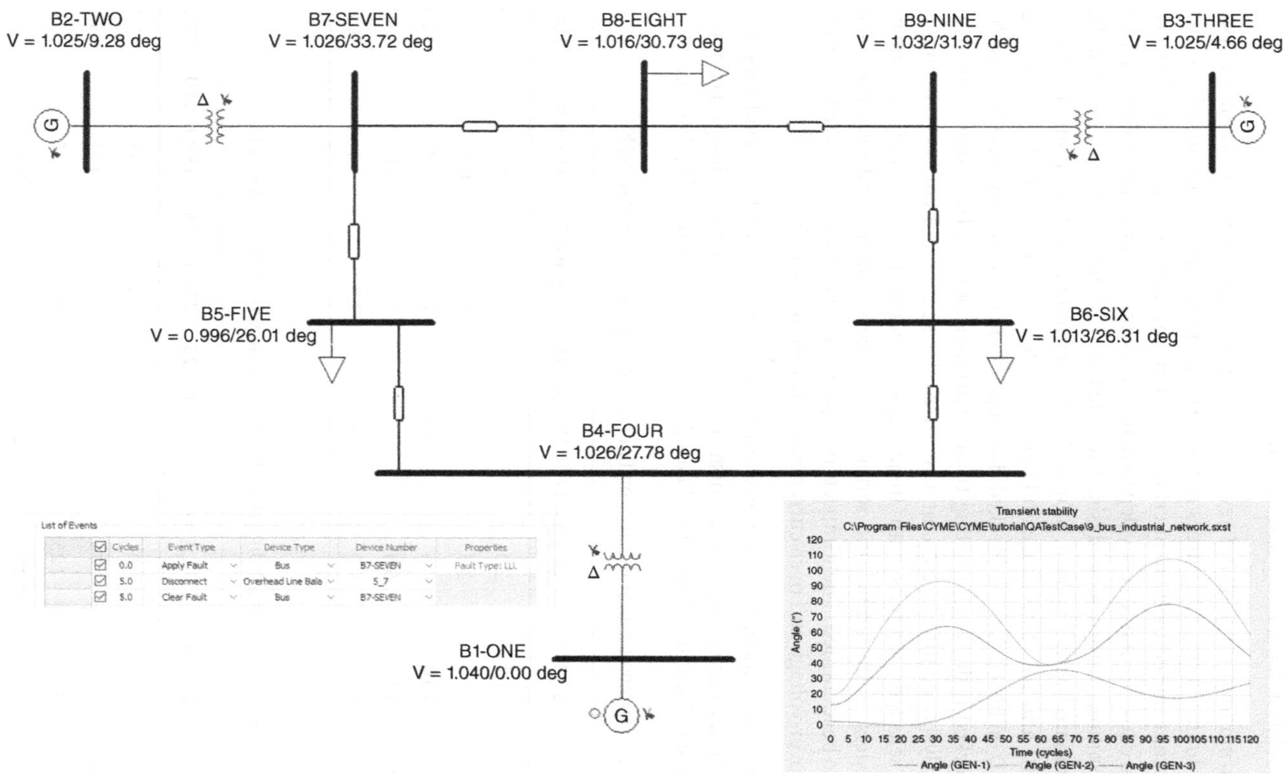

Figure 2.11 Sample transient stability analysis result display.

synchronous generators is plotted against each other to examine relative rotor angle deviation for stability assessment.

10) *Harmonic distortion and resonance analysis*. Harmonic distortion analysis evaluates harmonic sources in the system and the distortions in voltage and current caused by them at critical locations throughout the system. Network and equipment frequency dependent models and parameters shall be used for harmonic distortion analysis. The calculated harmonic distortions shall be compared with standard, or user defined harmonic content limits. Typically, THD and IHD for voltage and current at PCC with utility, voltage THD and IHD on buses shall be checked. Total current in the neutral wires shall also be checked if there are existing triplen harmonic source. Harmonic resonance analysis identifies resonant conditions by scanning frequency responses at selected locations with frequency-dependent network and equipment models. Both parallel and serious resonance situations and locations shall also be identified. Their impacts on equipment and system safety in the forms of over-voltage and over-current shall be carefully studied.

For I&C power systems, acceptance criteria for harmonic analysis results regarding harmonic current and voltage distortions at the PCC of the utility grid connection are specified in Tables 2.1–2.4 based on IEEE 519 Standard [7]: Where,

Table 2.1 Harmonic current distortion limit for voltage range 120 V to 69 kV.

I_{sc}/I_L	$3 \leq h < 11$	$11 \leq h < 17$	$17 \leq h < 23$	$23 \leq h < 35$	$35 \leq h \leq 50$	TDD
<20	4.0	2.0	1.5	0.6	0.3	5.0
20 < 50	7.0	3.5	2.5	1.0	0.5	8.0
50 < 100	10.0	4.5	4.0	1.5	0.7	12.0
100 < 1000	12.0	5.5	5.0	2.0	1.0	15.0
>1000	15.0	7.0	6.0	2.5	1.4	20.0

Table 2.2 Harmonic current distortion limit for voltage range 69 kV to 161 kV.

I_{sc}/I_L	$3 \leq h < 11$	$11 \leq h < 17$	$17 \leq h < 23$	$23 \leq h < 35$	$35 \leq h \leq 50$	TDD
<20	2.0	1.0	0.75	0.3	0.15	2.5
20 < 50	3.5	1.75	1.25	0.5	0.25	4.0
50 < 100	5.0	2.25	2.0	0.75	0.35	6.0
100 < 1000	6.0	2.75	2.5	1.0	0.5	7.5
>1000	7.5	3.5	3.0	1.25	0.7	10.0

Table 2.3 Harmonic current distortion limit for voltage range greater than 161 kV.

I_{sc}/I_L	$3 \leq h < 11$	$11 \leq h < 17$	$17 \leq h < 23$	$23 \leq h < 35$	$35 \leq h \leq 50$	TDD
<25	1.0	0.5	0.38	0.15	0.1	1.5
25 < 50	2.0	1.0	0.75	0.3	0.15	2.5
≥50	3.0	1.5	1.15	0.45	0.22	3.75

Table 2.4 Harmonic voltage distortion limit.

Bus voltage at PCC	IHD (%)	THD (%)
$V \leq 1.0\,kV$	5.0	8.0
$1\,kV < V \leq 69\,kV$	3.0	5.0
$69\,kV < V \leq 161\,kV$	1.5	2.5
$161.001\,kV < V$	1.0	1.5[a]

I_{sc}/I_L is the ratio of short-circuit current in percent available at PCC to the maximum fundamental load current.

TDD is total demand distortion (TDD) defined as a ratio of the root mean square of the harmonic content, considering harmonic components up to the 50th order (order greater than 50 may be included when necessary) and specifically excluding interharmonics, expressed as a percent of the maximum demand current.

In addition to harmonic distortions, parallel resonance and capacitor protections shall also be enforced. IEEE Standard 18 and relevant literature [8, 9] specify voltage limitations to shunt capacitor banks in terms of equipment ratings or the maximum rating permitted as following:

a) continuous operating voltage ≤110% of the rated voltage
b) rms crest voltage ≤1.2 times the rated rms voltage
c) kvar ≤ 135% of the rated kvar
d) current ≤ 180% of the rated rms currents

Where continuous operating voltage, rms crest voltage, and current should include harmonic components, especially resonance voltage.

Remedy or mitigation actions if needed shall be recommended and verified including but not limited to installing passive and active filters, requesting multi-pulse drives (e.g., 12-pulse or 18-pulse or higher), specifying special phase shifting transformer, sizing line chocks (reactors), relocating harmonic sources, retrofitting or upgrading equipment, etc. The overall objective is to maintain power quality within acceptable range without damaging equipment or interrupting the processes.

A sample harmonic distortion analysis result is shown in Figure 2.12 in which a harmonic source is connected at bus ID RECT to inject harmonic currents into the network. The table presents calculated harmonic indices of phase A voltage at bus RECT-A, and the plot gives harmonic spectrum verse frequency.

11) *Other important analyses*. Depending on the system configuration, load characteristics, and operation requirements, there may be special and dedicated studies needed for different purposes and applications. Those studies include but not limited to: economic analysis for energy efficiency and power losses, optimal power flow for a coordinated solution to achieve minimum cost objectives by leveraging available power system control devices and parameters while meeting equipment and operational constraints, voltage stability analysis to identify locations potentially causing voltage collapse and the associated transferring power and load power thresholds, emergency and backup power analysis for power source adequacy, and system restart and black start analysis for assessing island start capability, etc.

I&C power system detailed design and studies shall follow applicable industrial standards and corporate manuals to ensure the final design can meet operation requirements under normal, special and critical conditions.

2.2.4 Multiphase Analysis

2.2.4.1 System Analysis Phases

I&C power system analysis is carried out along with the project progress. In practice, I&C power system project is always scheduled into multiple phases or stages which are signified by starting and completing system design, ordering and purchasing equipment, installatiing and commissioning the equipment and the system. Typical project stages are:

1) *For review stage (FR)*. In this stage, system analyses are conducted based on the electrical system database that is updated in front end engineering design (FEED) stage. Completion of FR will move the project into a design stage.
2) *For design stage (FD)*. In this stage, system analyses are conducted based on the electrical system database that is updated in completion for review stage.
3) *For construction stage (FC)*. In this stage, system analyses are conducted based on the finalized electrical system database with completed vendor data.

2.2.4.2 Database Evolution

An important thing to understand when planning time and resources to perform system analysis is that the system database evolves when the project advances into a newer stage. During each stage, the system database is updated based on

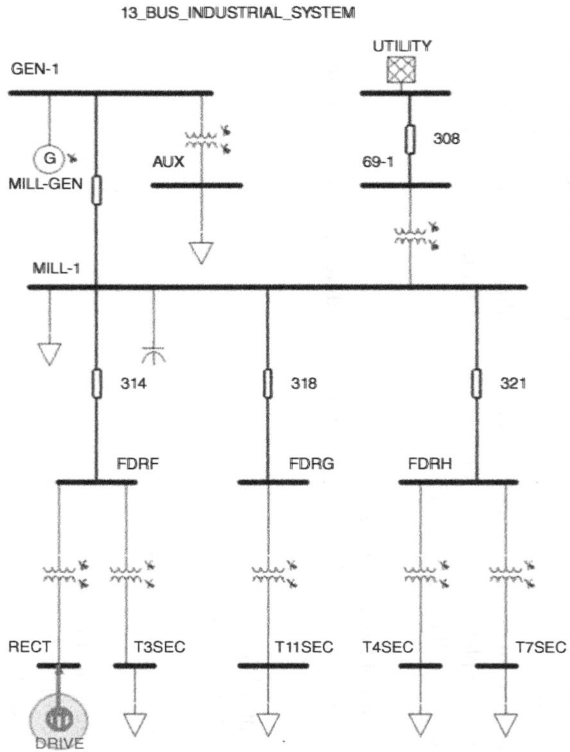

Node ID	Freq. (Hz)	Real V (V)	Imag V (V)	Modulus V (V)	Angle (deg)	IHD (%)	THD (%)	KVT (kV)	TIF
RECT-A	60.00	269.5534	-22.0857	270.4567	-4.68	0.000	7.831	55.97	206.30
RECT-A	300.00	12.7700	1.5200	12.8600	6.79	4.755	7.831	55.97	206.30
RECT-A	420.00	7.3790	-9.2710	11.9300	-51.78	4.411	7.831	55.97	206.30
RECT-A	660.00	-2.0420	-6.7540	7.0560	-106.82	2.609	7.831	55.97	206.30
RECT-A	780.00	-5.0610	-3.2170	5.9970	-147.56	2.217	7.831	55.97	206.30
RECT-A	1020.00	-1.9890	3.2990	3.8520	121.09	1.424	7.831	55.97	206.30
RECT-A	1140.00	1.1300	2.9030	3.1160	68.73	1.152	7.831	55.97	206.30
RECT-A	1380.00	1.8820	-1.7420	2.5650	-42.79	0.948	7.831	55.97	206.30
RECT-A	1500.00	-0.2292	-2.5440	2.5540	-95.15	0.944	7.831	55.97	206.30
RECT-A	1740.00	-2.4020	0.4948	2.4520	168.36	0.907	7.831	55.97	206.30
RECT-A	1860.00	-1.1900	1.9570	2.2910	121.30	0.847	7.831	55.97	206.30
RECT-A	2100.00	1.6890	0.7233	1.8380	23.18	0.680	7.831	55.97	206.30
RECT-A	2220.00	1.4650	-0.8192	1.6780	-29.22	0.620	7.831	55.97	206.30

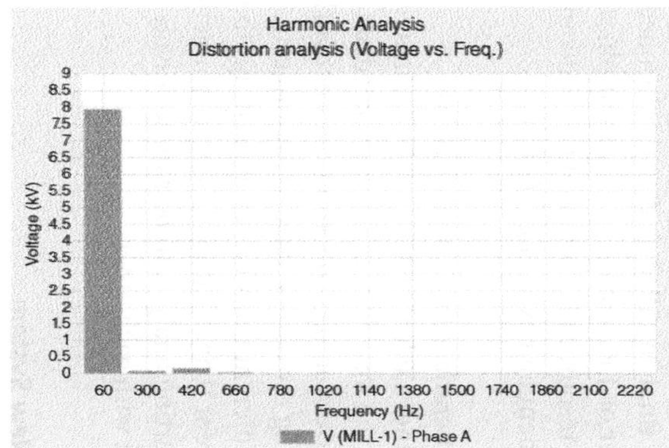

Figure 2.12 Sample harmonic distortion analysis result display.

equipment procurement and delivery, and one set of complete studies based on the design specifications and project scope shall be run to produce new reports for owners or engineering, procurement, and construction (EPC) companies to review. At end of the project, an as-built database shall be fully validated with the final installed and commissioned equipment data and submitted to the owner or EPC companies. A multiphase project can last for months or even years, so resource allocation and data documentation can be crucial for smooth management and execution of projects.

2.2.4.3 Iterations for System Analysis

It is also inevitable that each analysis can iterate multiple times. For example, iterations of the load flow analysis are needed to determine optimal transformer tap settings to cover load conditions and network impedances which cannot be finalized during the earlier stage of the project. Also, the motor starting analysis may require different transformer tap settings to remedy starting voltage drops which are different than what load flow recommends. This requires both load flow and motor starting analysis to be run iteratively for multiple phases and cross check the result with each other.

2.2.5 New System

A new I&C power system is referring to a greenfield project in which a new system infrastructure will be designed and built including power sources, distribution network, loads, protections, etc. The new installation of an I&C power system requires comprehensive analyses to ensure system operability, reliability, efficiency, and safety. The complete design and study cycles including conceptual design and detailed design will be carried out.

2.2.5.1 Types of System Analysis

Studies that should be included for a newly designed I&C power system are listed below:

1) Load profile analysis
2) Load flow analysis
3) Equipment sizing
4) Power factor correction
5) Short-circuit analysis
6) Voltage drop analysis (*)
7) Protective device coordination
8) Motor starting analysis
9) Harmonic distortion analysis (*)

10) Transient stability analysis (*)
11) Emergency and backup power study (*)

*Depends on system applicability and equipment availability.

2.2.5.2 Data Considerations

Validity and accuracy of system studies and analysis for I&C power systems at any time heavily depend on the system model and data fidelity. Determining and providing system and equipment data for studies and analysis thus is a critical part for new I&C power systems and data validation has the very high priority. Initially, only typical data or historical data are available which are probably used for conceptual design. Along with the project progress, per study recommendations and procurement preferences, vendor data become available for some equipment. The project analysis database will be updated iteratively to reflect inputs from the study team and procurement team. System study and analysis tasks need to be organized to meet the project cycle and make sure studies are performed on the latest database and the new recommendations are presented. The final project analysis database should be based on as-build, taking from equipment nameplate, corresponding datasheet and filed commissioning test if necessary.

2.2.5.3 Operation Requirements and Guidelines

In addition to database evolution throughout the complete project cycle, attention must also be given to the system operation requirements. There are two major stakeholders for the project technical requirements, one is from the facility owner(s), and another is from the utility or regional power provider. The facility owner(s) will decide the final loads and processes which could affect the design concept and study guidelines. On the utility side, design and study recommendations need to comply with the utility regulation requirements.

2.2.6 Existing System

2.2.6.1 Needs for Existing System Analysis

Electrical system analysis and studies are an integral part of system design, operation, and maintenance. They shall be performed at the design stages for new projects, as well as conducted on a regular basis for existing systems. When the existing system undergoes new modifications in system configuration, changes in generation or load, or updates in operations, system analysis and studies shall be updated. In addition, system studies shall be redone every five years. ANSI/NETA MTS-2023 Standard for Maintenance Testing Specifications for Electrical Power Equipment and Systems specifies power system study frequency in Table 2.5 [10].

Table 2.5 Frequency of power system studies.

Circumstance	One-line diagram update	Short-circuit study	Coordination study	Incident energy analysis	Load flow study
New installation or system modification	X	X	X	X	X
Annual data verification	X	X	X	X	X
Change in utility or source	X	X	X	X	X
Change in system impedance, configuration, or loading	X	X	—	X	—
Change in protection devices or settings	X	—	X	X	X
5 years since last update	X	X	—	X	—

Source: Adapted from ANSI/NETA [10].

2.2.6.2 Some Examples for Existing System Analysis

Existing I&C power systems may need to conduct system analysis, either focusing on some particular areas, or a complete model and data revalidation and analysis.

Examples of when load flow analysis is needed including system performance improvements or requirements compliance are [3]:

1) Power factor correction
2) Power flow operating requirements handed down by regulating authorities (reactive power demand based on voltage fluctuations at PCC)
3) Voltage profile improvement by reactive power support (i.e., the use and selection of static var. compensators, capacitor banks, automatic capacitor switching, and adjustable speed drives to reduce reactive power demand)
4) Reduction of real power losses in branch elements like cables and transformers by better selection of transformer tap positions, better current limiting device selection

Other examples of initiating special or complete analysis of an existing system include but not limited to:

1) Protection coordination miscoordination
2) Failure in fault current interruption
3) Motor starting fails
4) Unacceptable voltage drops during motor starting
5) Excessive harmonic distortions
6) Harmonic resonances

7) Oscillations related to synchronous machines
8) Low frequency or voltage during transients

Study and analysis of existing systems or brownfield projects do not need conceptual design. The study and analysis will assess the existing electrical infrastructure to ensure the system still meets operability, reliability, efficiency, and safety standards. The study may be based on the scheduled activity, or need for troubleshooting, or requirement for new standard compliance, or plan to prepare for potential system modification or upgrade.

2.2.6.3 Types of Existing System Analysis

Studies for the existing system really depend on the rationale of the project; therefore, specific studies that are scoped could be very different from project to project. Nevertheless, following standard studies are within the consideration:

1) Load profile analysis (*)
2) Load flow analysis (*)
3) Equipment sizing (*)
4) Power factor correction (*)
5) Short-circuit analysis (*)
6) voltage drop analysis (*)
7) Protective device coordination (*)
8) Motor starting analysis (*)
9) Harmonic distortion analysis (*)
10) Transient stability analysis (*)
11) Emergency and backup power study (*)

*Depends on the project scope and goals, as well as the equipment availability.

2.2.6.4 Data Considerations

Data for existing system normally should be available in one of the following formats:

1) Existing system model
2) System owner(s)/engineer(s) database
3) Original datasheet
4) Combination of the above

2.2.6.5 Model Validation

When performing studies and analysis of existing systems, opportunities should be seriously taken to verify and validate the complete system model data. There are many ways to perform model data validation. It can be done typically in a couple of practical approaches:

1) Through benchmark study results
2) Per vendor factory acceptance test (FAT) and site acceptance test (SAT) results
3) According to on-line measurements and state estimation results
4) By field measurements
5) From fault or event recording data

Validated data should be kept in the database and record to be used for studies until the next time when data are updated and validated again.

2.2.6.6 Operation Requirements and Guidelines

Studies and analysis for the existing I&C power system should also check and try to comply with operation requirements by the facility owner(s) and interconnection guidelines from the local utility. Copies of single-line diagrams and system study data should be given to the facility maintenance department. To ensure efficient and safe system operation, it is critical for the maintenance department to keep the single-line diagrams current and discusses significant changes with the facility engineering department or consulting electrical engineer. It should be noted, however, that the information required for system studies is highly specialized, and outside help might be necessary.

2.2.7 System Expansion

When an existing I&C power system is expanded to add new facilities and processes, often with a significant number of new equipment, the project most likely will need to go through both conceptual design and detailed design too. System expansion involves assessing and planning the growth of electrical infrastructure to meet increasing demand or accommodate new facilities. This process requires a thorough understanding of the existing power system, as well as considerations for future requirements. The study and analysis scope may only cover the added system, but the study model must be the entire system including the existing system. Sometimes, the study scope could be the entire system too. Depending on whether the added system will contain power sources, distribution network, loads, and protections, etc., study and analysis related to system overall operability, reliability, efficiency and safety should be covered.

2.2.7.1 Types of System Analysis

Studies that should be included for an expanded I&C power system are basically the same as for a newly designed I&C power system, subject to the actual system expansion scope and equipment added:

1) Load profile analysis
2) Load flow analysis
3) Equipment sizing
4) Power factor correction
5) Short-circuit analysis
6) voltage drop analysis (*)
7) Protective device coordination
8) Motor starting analysis (*)
9) Harmonic distortion analysis (*)
10) Transient stability analysis (*)
11) Emergency and backup power study (*)

*Depends on system applicability and equipment availability.

2.2.7.2 Data Considerations

Data for studying and analyzing an expanded system should include both the existing system and the newly added system. This is a very challenging task. Existing system data should be validated before being used as a foundation for the integrated system database, and new system data will go through the same cycles as for a newly designed system study and analysis, i.e., typical and historical data in the conceptual design phase, vendor data in the detailed design phase, and as-build data in the final study and analysis phase. Iterations in database and studies are expected along with the project progress in procurement and vendor deliverable.

2.2.7.3 Operation Requirements and Guidelines

Studies and analysis of I&C power systems should also consider system operation requirements from both the facility owner(s) and interconnection guidelines from the local utility. In the latter case, it is necessary since the changes inside the I&C power system may vary the active and reactive power demands, fault current duties, harmonic injections and distortions at the PCC which will infect the utility reliability and protection.

2.2.7.4 Issues Related to the Existing System

When there are code or guideline violations found in the existing system, it is needed to consult with the facility owner(s) whether the study and analysis scope should be extended to the entire system, if the initial scope of work (SOW) only covers the added system. In case the SOW is restricted to the added system, findings and recommendations related to the existing system should still be clearly made in the study reports. Reference [11] deals with a complete study of a system with major expansions and limited SOW to the added system only. The system high level single-line diagram is shown in Figure 2.13, where the dash-bounded sections are the new expansion.

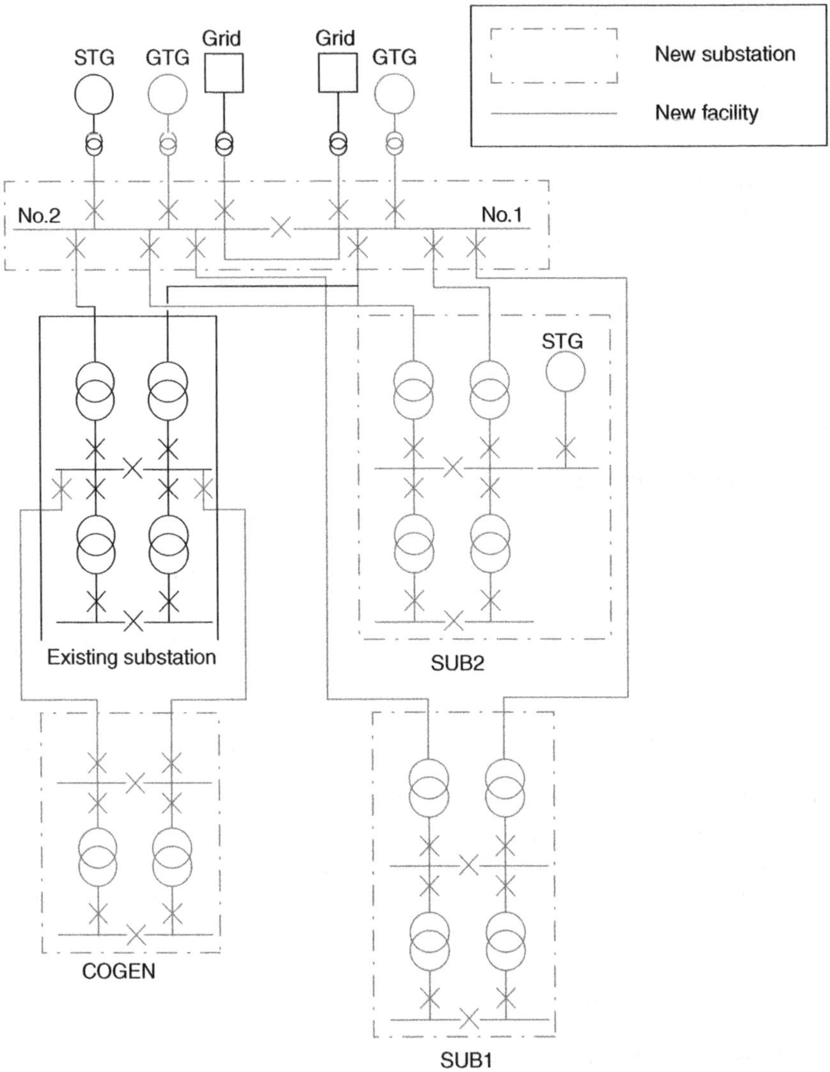

Figure 2.13 Sample of a major I&C power system expansion single-line diagram. Source: Dai [11]/with permission of IEEE.

2.3 Scenarios

Study and analysis cases are based on real system scenarios which are selected from designed normal operation conditions, start-up, shut-down, maintenance and other special or abnormal modes, major N-1 contingencies, and recommendations by industry standards. For each scenario, appropriate power system studies and analysis such as load flow, short-circuit, protection and device coordination, motor starting, harmonic and transient stability will be applied to model and simulate the system characteristics and responses under the given operating conditions. These analyses will provide data and information for engineers to enhance and add protections to the system to ensure stable and safe operations. Selected scenarios should be realistic, representative, and concise since studies and analysis are very time consuming, thus similar, redundant and insignificant scenarios will waste valuable engineering time and project budget. The major variation in a study scenario includes following parameters:

1) System configuration
2) Protective device position (close or open)
3) Equipment status (on or off)
4) Initial loading
5) Initial generation
6) Initial power import/export
7) Initial voltage profile

Generally, scenario selections can be categorized based on study and analysis goals, such as planning, checking normal operation, assessing contingencies, and optimizing controls. Scenarios selection guidelines are described in the following for each of the major study types in different bases.

2.3.1 Planning Based

The main objective of I&C power system planning is to ensure designing of an electrical infrastructure that can provide adequate power capacity to the loads, operate reliably and economically efficiently, comply with grid regulatory requirements, and have scalability to grow for future expansions. Planning based study scenarios should be selected to assist planners and engineers in making strategic decisions to meet current and future electricity demand while considering various technical, economic, environmental, and social factors. Aspects for each study and analysis by considering planning process are mainly focused on conceptual level for load flow, short-circuit and motor starting analyses.

2.3.1.1 Load Flow Analysis

For load flow analysis, the design parameters are transformer, line and cable, current limiting reactor (CLR), and load feeder sizes and ratings, power factor correction capacitor bank and other shunt compensation device sizes, whereas the adjustable parameters are generator voltage setting, generator active and reactive power generations, transformer tap settings, capacitor and shunt compensation device settings, if they are not pre-fixed. System configurations and load center locations are also to be finalized if the system allows us to do so. Load flow analysis scenarios shall be designed to provide information for the following investigations and decisions objectives:

1) Electrical load requirements of the processes by identifying and considering peak, average, minimum, and future projected loads
2) Load types, characteristics, and diversities
3) Voltage variations
4) Power factor values
5) Active and reactive power generation adequacy
6) Generator voltage regulations
7) Transformer tap positions
8) Capacitor and shunt compensation settings
9) Equipment sizes and ratings
10) System configurations

2.3.1.2 Short-Circuit and Fault Analysis

For short-circuit and fault analysis, the focus is to calculate the highest possible fault duties at key locations in order to evaluate the impact on the power system, select and verify or size circuit breaker capabilities and fuse ratings to interrupt the faults and ensure the system safety. Fault current contributions from sources include utilities, on-site generators, motors and other energy generation and storage devices, as well as system configurations will affect the results. Therefore, short-circuit and fault study and analysis scenarios should consider the following conditions:

1) Condition corresponding to the maximum utility short-circuit MVA (Mega Volt Amp) at PCC
2) Condition corresponding to the maximum generator fault current contributions
3) Condition corresponding to the maximum fault current contributions from motors
4) Condition corresponding to the maximum fault current contributions from energy storage devices
5) Condition corresponding to the highest pre-fault voltage
6) System configurations corresponding to the above conditions

2.3.1.3 Motor Starting Analysis

Usually only static motor starting analysis is required during the planning stage. This analysis models the starting motor as an equivalent load with its reactive power set to the in-rush value (around five to seven times of the rated value) and the active power to the value depending on the process requirement (from 0 to 100% of the rated loading). The scenarios normally consider:

1) Start the largest motor at each substation or bus

2.3.2 Normal Operation Based

System study and analysis for I&C power systems under normal operating conditions is crucial to ensure system reliability, efficiency and safety. The scenarios should be designed and selected to fully confirm and/or update operation procedures, identify and remedy violations, verify and/or adjust protections, and validate operational schedules and programs. The scenarios should also consider some special situations when the system is started up and brought up to the normal operation mode, or transitioned from the normal operation mode to other modes.

All scenarios on the normal operation base together with the study output should be documented as benchmark results to be used as references for checking and comparison for other condition-based studies and analysis.

2.3.2.1 Load Flow Analysis

Load flow analysis for normal operation systems is to evaluate system performance including voltage drop, reactive power compensation, load balance, equipment loading, generation support capability, etc. The analysis scenarios shall cover the following cases in Table 2.6.

2.3.2.2 Short-Circuit and Fault Analysis

Short-circuit analysis for normal operation system should be designed and selected to thoroughly evaluate the impact of short-circuit at different locations in the system and assess the adequacy of protection devices including breakers and fuses under the highest fault current duties to ensure device ratings are not exceeded, as well as to protect no thermal and mechanical damages to the equipment which are exposed to the fault currents. The analysis scenarios shall cover the following cases in Table 2.7.

2.3.2.3 Protection and Device Coordination

Protection and device coordination will generate TCCs for each protection zone, based on three-phase and single-line-to-ground fault currents at the terminal of the zone. Zones are composed of busbars, transformers, cables, lines, generators, motors and other electrical distribution equipment on the fault path and co-located in the same area. Principles for fault protections are (1) to minimize the down time during faults; (2) to minimize the number of equipment due to the fault interruption and isolation. This study should check, evaluate and adjust protection device parameters and settings for relays, breakers, current transformers (CTs),

Table 2.6 Load flow scenarios for normal operation-based analysis.

	Objective	Condition	Importance
(a)	• Voltage regulation • Active power balance • Reactive power support • Power factor requirement • Equipment overloading • Overall system performance	• Normal operation • Minimum generation • Maximum operation loads • Configuration i ($i = 1, \ldots, n$)	High
(b)	• Voltage regulation • Active power balance • Reactive power support • Power factor requirement • Equipment overloading • Overall system performance	• Normal operation • Normal generation • Maximum operation loads • Configuration i ($i = 1, \ldots, n$)	High
(c)	• Voltage regulation • Active power balance • Equipment overloading • Overall system performance	• Essential operation • Essential generation • Emergency loads • Configuration i ($i = 1, \ldots, n$)	High
(d)	• Voltage regulation • Active power balance • Reactive power support • Equipment overloading • Overall system performance	• Special operation • Normal and minimum generation • Corresponding loads • Special configurations (such as at start up, partial or full shut down, maintenance modes, etc.)	Medium

Notes: Configurations can represent utility connections, bus-tie status, equipment maintenance conditions, etc.

potential transformers (PTs) and other faulty current sensors to ensure proper coordination between devices to meet those principles. The protection and device coordination scenarios shall consider covering the following cases in Table 2.8.

2.3.2.4 Motor Starting Analysis

Motor starting analysis is brought in to check and verify that designed processes driven by motors are not interrupted by system voltage variations, and motors can be successfully started and accelerated to the rated speed under qualified conditions. Motor starting analysis at the operation-based stage will use dynamic models for motors including their loads to obtain more accurate and realistic

Table 2.7 Short-circuit scenarios for normal operation-based analysis.

	Objective	Condition	Importance
(a)	• Calculate fault current duties: • Momentary peak • Interruption • Steady state maximum • Check protective device capabilities: • Closing and latching • Interruption (breaking) • Long time withstands	• Normal operation • Maximum generation • Maximum motor load • Configuration i ($i = 1, \ldots, n$) • Faults at each individual bus location: • Three-phase • Line-to-line • Line-to-line-to-ground • Single-line-to-ground	High
(b)	• Calculate fault current duties: • Steady state minimum • Check protective device capabilities: • Long time sensitivity	• Normal operation • Minimum generation • Minimum motor load • Configuration i ($i = 1, \ldots, n$) • Faults at each individual bus location: • Three-phase • Line-to-line • Line-to-line-to-ground • Single-line-to-ground	Medium
(c)	• Calculate fault current duties: • Momentary peak • Interruption • Steady state • Check protective device capabilities: • Closing and latching • Interruption (breaking) • Long time withstand	• Momentary operation • Essential generation • Essential motor load • Essential configuration • Faults at each individual bus location: • Three-phase • Line-to-line • Line-to-line-to-ground • Single-line-to-ground	High
(d)	• Calculate fault current duties: • Momentary peak • Interruption • Steady state • Check protective device capabilities: • Closing and latching • Interruption (breaking) • Long time withstand	• Momentary operation • Emergency generation • Emergency motor load • Emergency configuration • Faults at each individual bus location: • Three-phase • Line-to-line • Line-to-line-to-ground • Single-line-to-ground	High

Notes: Configurations can represent incoming line circuit breaker open or close status, high, middle and low voltage bus tie switchgear open and close conditions, etc.

Table 2.8 Protection and device coordination scenarios for normal operation-based analysis.

	Objective	Condition	Importance
(a)	• Set and coordinate protection devices in the zone to meet requirements of: • Fast clearing time • Selectivity • Time delay	• Over current protection zone for each potential fault location • Three-phase and single-phase faults	High
(b)	• Set and coordinate protection devices in the zone to meet requirements of: • Avoid fault current energy to exceed thermal damage curves of the equipment in the zone • Avoid current interruption to motor and transformer inrush • Avoid load current exceeding the limits while allow acceptable load current without interruption	• Over current protection zone for each potential fault location • Equipment thermal damage curves • Motor inrush current starting curves and transformer inrush current energizing point • Maximum load current	High

results. For the cases that motors failed to start or accelerate, starting devices should be designed and verified. Consequently, motor starting analysis scenarios shall include the cases in Table 2.9.

2.3.2.5 Harmonic Analysis

Harmonic analysis at the operation stage will model and simulate all existing harmonic sources and investigate harmonic voltage and current distortion that can affect the performance of sensitive loads and devices, identify parallel and serious harmonic resonances that can cause over-voltage or over-current damages. Harmonic mitigation and remedies should be proposed, designed and verified. Harmonic study scenarios should cover the following cases in Table 2.10.

2.3.2.6 Transient Stability Analysis

Transient stability study at the operation stage needs to capture system response to sudden disturbances that potentially jeopardize synchronous machine rotor angle stability and cause voltage and frequency oscillation continuously and out of ranges. It also proposes, designs and verifies solutions to ensure the system could recover from disturbances and maintain stable operation with all synchronous machines in synchronism. Thus, attention must be given to select and define disturbances in the system such as faults, generation impacts, load impacts, sudden configuration changes, etc. The typical transient stability study scenarios shall include the cases in Table 2.11.

Table 2.9 Motor starting scenarios for normal operation-based analysis.

	Objective	Condition	Importance

(a)	• Check voltage drops at:
	  • Starting motor terminal
	  • Running motor bus
	  • Generator bus
	• Ensure inrush reactive power demand
	• Validate motor starting time
	• Evaluate voltage regulation capability (generator AVR, transformer tap)
	• Assess staring device effectiveness (if any)
		• Normal operation
		• Normal generation
		• Maximum operation load
		• Configuration i ($i = 1, \ldots, n$)
		• Starting motors for:
		  • The largest motor on each bus or substation at a time
		  • Batch starting per starting schedule
		  • With or without starting device per design
			High

(b)	• Check voltage drops at:
	  • Starting motor terminal
	  • Running motor bus
	  • Generator bus
	• Ensure inrush reactive power demand
	• Validate motor starting time
	• Evaluate voltage regulation capability (generator AVR, transformer tap)
	• Assess staring device effectiveness (if any)
		• Normal operation
		• Minimum generation
		• Maximum operation load
		• Configuration i ($i = 1, \ldots, n$)
		• Starting motors for:
		  • The largest motor on each bus or substation at a time
		  • Batch starting per starting schedule
		  • With or without starting device per design
			High

(c)	• Check voltage drops at:
	  • Starting motor terminal
	  • Running motor bus
	  • Generator bus
	• Ensure inrush reactive power demand
	• Validate essential motor starting time
	• Evaluate voltage regulation capability (generator AVR, transformer tap)
	• Assess staring device effectiveness (if any)
		• Essential operation
		• Essential generation
		• Essential operation load
		• Essential configuration
		• Starting essential operation motors
			High

(d)	• Check voltage drops at:
	  • Starting motor terminal
	  • Running motor bus
	  • Generator bus
	• Ensure inrush reactive power demand
	• Validate emergency motor starting time
	• Evaluate voltage regulation capability (generator AVR, transformer tap)
	• Assess staring device effectiveness (if any)
		• Emergency operation
		• Emergency generation
		• Emergency operation load
		• Emergency configuration
		• Starting emergency operation motors
			High

Notes: Configurations can represent incoming line circuit breaker open or close status, bus tie open and close conditions, equipment in maintenance conditions, etc.

Table 2.10 Harmonic scenarios for normal operation-based analysis.

	Objective	Condition	Importance
(a)	• Evaluate voltage distortions at PCC, critical and sensitive buses • Evaluate current distortions at PCC • Assess neutral currents • Check equipment overloading • Identify parallel resonance conditions • Identify serious resonance conditions • Recommend remedies for harmonic distortions (such as filter, chock, device upgrade, K-transformer, transformer phase shift, harmonic source relocation, etc.) • Recommend remedies for parallel and serious resonance damages (such as filter, device upgrade, capacitor bank relocation, harmonic source relocation, etc.)	• Normal operation • Normal generation • Maximum operation load • Configuration i ($i = 1, \dots, n$) • All harmonic sources included • Capacitor banks at the maximum settings	High
(b)	• Evaluate current distortions at PCC • Assess neutral currents • Identify parallel resonance conditions • Identify serious resonance conditions • Recommend remedies for harmonic distortions (such as filter, chock, device upgrade, K-transformer, transformer phase shift, harmonic source relocation, etc.) and coordinate with the base (normal) case solutions • Recommend remedies for parallel and serious resonance damages (such as filter, device upgrade, capacitor bank relocation, harmonic source relocation, etc.) and coordinate with the base (normal) case solutions	• Essential operation • Essential generation • Essential operation load • All involved harmonic sources included • All involved capacitor banks at the maximum settings	Medium
(c)	• Evaluate current distortions at PCC • Assess neutral currents • Identify parallel resonance conditions • Identify serious resonance conditions • Recommend remedies for harmonic distortions (such as filter, chock, device upgrade, K-transformer, transformer phase shift, harmonic source relocation, etc.) and coordinate with the base (normal) case solutions • Recommend remedies for parallel and serious resonance damages (such as filter, device upgrade, capacitor bank relocation, harmonic source relocation, etc.) and coordinate with the base (normal) case solutions	• Emergency operation • Emergency generation • Emergency operation load • All involved harmonic sources included • All involved capacitor banks at the maximum settings	Medium

Notes: Configurations can represent incoming line circuit breaker open or close status, high, middle and low voltage bus tie switchgear open and close conditions, equipment maintenance conditions, etc.

Table 2.11 Transient stability scenarios for normal operation-based analysis.

	Objective	Condition	Importance
(a)	• Assess rotor angle stability • Assess voltage deviations • Assess frequency variations • Recommend protections and mitigation	• Normal operation • Normal generation • Normal operation load • Configuration i ($i = 1, \ldots, n$) • Main generator trip	High
(b)	• Assess rotor angle stability • Assess voltage deviations • Assess frequency variations • Evaluate system oscillations • Recommend protection and mitigation	• Normal operation • Minimum generation • Normal operation load • Configuration i ($i = 1, \ldots, n$) • Main generator trip	High
(c)	• Evaluate system oscillations • Recommend protection and mitigation	• Normal operation • Normal generation • Normal operation load • Configuration i ($i = 1, \ldots, n$) • The largest load trip	High
(d)	• Evaluate system oscillations • Recommend protection and mitigation	• Normal operation • Minimum generation • Normal operation load • Configuration i ($i = 1, \ldots, n$) • The largest load trip	High

(e)	• Evaluate rotor angle stability
	• Assess voltage deviations
	• Assess frequency variations
	• Evaluate system oscillations
	• Estimate critical fault clearing time (CFCT)
	• Recommend protection and mitigation
		• Normal operation
		• Normal generation
		• Normal operation load
		• Configuration i ($i = 1, \ldots, n$)
		• Three-phase faults at:
		  • Generator terminal bus
		  • High voltage switchgear
			High

(f)	• Evaluate rotor angle stability • Assess voltage deviations • Assess frequency variations • Evaluate system oscillations • Estimate critical system separation time (CSST) • Recommend protection and mitigation	• Normal operation • Normal generation • Normal operation load • Configuration i ($i = 1, \ldots, n$) • Three-phase faults at: • Utility connection	High

Table 2.11 (Continued)

	Objective	Condition	Importance
(g)	• Evaluate rotor angle stability • Assess voltage deviations • Assess frequency variations • Evaluate generator governor and excitation responses and controllability • Recommend load shedding protection scheme	• Normal operation • Minimum generation • Maximum operation load • Configuration i ($i = 1, \ldots, n$) • Planed and forced islanding	High
(h)	• Evaluate rotor angle stability • Assess voltage deviations • Assess frequency variations • Evaluate generator governor and excitation responses and controllability • Recommend load shedding protection scheme	• Normal operation • Minimum generation • Normal operation load • Configuration i ($i = 1, \ldots, n$) • Planed and forced islanding	High

Notes: Configurations can represent incoming line circuit breaker open or close status, bus tie switchgear open and close conditions, equipment maintenance conditions, etc.

2.3.3 Contingency Based

Contingency based study scenarios for I&C power systems focus on incidents particularly in the power distribution network and sources, such as potential equipment failures, utility grid outages, natural disasters and other unexpected disturbances, to assess if the system is sustainable in the capability to remain healthy power delivery and operation. A base case analysis of load flow is needed as a benchmark to compare and access system changes during contingencies. This shall be available from the standard load flow results for the same system to be analyzed.

Contingency based study and analysis is also referred as steady-state security analysis for load flow analysis and dynamic security analysis for transient stability analysis which will need to consider the following scenarios.

2.3.3.1 Load Flow Analysis

Load flow analysis is a fundamental tool to study system status under contingency conditions, such as the loss of a generator, tripping of a transmission line, a transformer, or a load, or disconnection with the grid. These studies will alert engineers to conditions of equipment overloads, poor voltage levels, loss of powers to critical and essential loads. Load flow analysis can provide solutions to remedy the above issues. Load flow analysis also establishes an initial condition for further performing transient stability analysis under contingency conditions. Table 2.12 shows typical contingency based load flow analysis scenarios.

2.3.3.2 Transient Stability Analysis

Transient stability should be reassessed under credited contingency conditions for synchronous generator rotor angle stability, system voltage and frequency stability, and other vulnerability under the disturbances that can affect system normal operation. Scenarios for transient analysis on contingency based are the same as benchmarked cases with specific objectives as shown in Table 2.13.

2.3.4 Optimization Based

I&C power system optimization is basically looking for solutions that can reduce equipment cost, minimize equipment size, improve operation efficiency, enhance system safety, increase system reliability, while maintain the system operability and sustainability. Optimization based power system studies and analysis are primarily utilizing optimal power flow technologies to deal with steady-state optimal solutions. Optimal power flow is a special formulated optimization process to search for a solution to minimize a cost function or an objective function through physically adjustable controls with defined system constraints. The objective functions typically are the minimum losses of active or reactive power in the network, maximum voltage security, maximum line flow security, minimum energy cost, etc., or any combination of them. The constraints are fundamental power flow equations coupled with some other equipment and system constraints and bounds such as line and cable thermal capacity, bus voltage operating range, generator capacity, etc. Control variables are adjustable devices in the system including transformer taps, generator AVRs, capacitor and shunt compensator settings, etc. Short-circuit and protection coordination can be combined to achieve certain optimization objectives if treat protection device setting as adjustable control variables.

There are some advanced technologies available at research level handling transient stability constrained optimal power flow. This is not included in the scope of this book.

2.3.4.1 Optimal Power Flow Analysis

The study scenarios for optimal power flow analysis at optimization stage should be designed to calculate or decide system configuration, equipment rating, and control settings based on given objective function and system constraints. Table 2.14 shows conditions and objectives involved in optimal power flow analysis scenarios.

Table 2.12 Load flow scenarios for contingency-based analysis.

	Objective	Condition	Importance
(a)	• Evaluate system steady state stability and voltage stability • Examine running generator overloading conditions • Asses load shedding or other remedies measurements to protect the system operability	• N-1 contingency • Trip of a main power generation unit i supplying a significant portion of the load • $i = 1, \ldots,$ number of main power generation units	High
(b)	• Evaluate remaining equipment overloading condition • Evaluate voltage stability • Assess power flow and distribution • Recommend network reconfiguration if necessary • Evaluate backup protection for main equipment	• N-1 contingency • Trip of a main power distribution equipment (transformer, line, cable, …) j delivering a significant portion of the load • $j = 1, \ldots,$ number of main power distribution equipment	High
(c)	• Evaluate the impact on voltage sensitive equipment (motors, control devices, etc.) • Assess the responses of voltage regulation devices • Recommend enhanced voltage regulation and protection if needed	• N-1 contingency • Significant voltage step change from the utility grid k • $k = 1, \ldots,$ number of in-coming utility grid connection	Medium
(d)	• Evaluate the impact on system voltage and power flow • Assess the protection ability and backup adequacy ability to isolate faulty equipment	• N-1 contingency • Sudden failure of critical switchgear l to affect power flow path • $l = 1, \ldots,$ number of critical switchgears	High
(e)	• Evaluate system steady state stability • Assess voltage stability • Examine generator and main equipment overloading • Evaluate generator and main equipment backup protection • Recommend needed enhancement and protection including load shedding, network reconfiguration, fault isolation, etc.	• N-2 contingency • Trip of main power generation unit i and main power distribution equipment j in combination of sudden failure of critical switchgear l • $i = 1, \ldots,$ number of main power generation units, $j = 1, \ldots,$ number of main power distribution equipment, $l = 1, \ldots,$ number of critical switchgears	Medium
(f)	• Evaluate system steady state stability • Assess voltage stability • Examine generator and main equipment overloading • Evaluate generator and main equipment backup protection • Recommend needed enhancement and protection including load shedding, network reconfiguration, fault isolation, etc.	• N-1-1 contingency • Trip of main power generation unit i and main power distribution equipment j in combination of sudden failure of critical switchgear l • For critical systems • $i = 1, \ldots,$ number of main power generation units, $j = 1, \ldots,$ number of main power distribution equipment, $l = 1, \ldots,$ number of critical switchgears	Medium

Table 2.13 Transient stability scenarios for contingency-based analysis.

	Objective	Condition	Importance
(a)	• Assess synchronous machine rotor angle stability • Evaluate frequency variation and recovery in particular the frequency nadir and rote of change of frequency (RoCoF) • Evaluate voltage variation and recovery • Assess generator speed governor and excitation system responses • Review frequency and voltage protection performance • Asses dynamic load shedding remedy and recommend solutions	• N-1 contingency • Trip of a main power generation unit i supplying a significant portion of the load • $i = 1, \ldots,$ number of main power generation units	High
(b)	• Assess synchronous machine rotor angle stability • Evaluate frequency variation and recovery • Evaluate voltage variation and recovery • Assess generator speed governor and excitation system responses • Review frequency and voltage protection performance	• N-1 contingency • Trip of a main power distribution equipment (transformer, line, cable, …) delivering a significant portion of the load • $j = 1, \ldots,$ number of main power distribution equipment	High
(c)	• Evaluate voltage variation and recovery • Assess generator excitation system responses • Review voltage protection performance	• N-1 contingency • Significant voltage step change from the utility grid k • $k = 1, \ldots,$ number of in-coming utility grid connection	Medium
(d)	• Assess synchronous machine rotor angle stability • Evaluate frequency variation and recovery • Evaluate voltage variation and recovery • Assess generator speed governor and excitation system responses	• N-1 contingency • Sudden failure of critical switchgear l to affect power flow path • $l = 1, \ldots,$ number of critical switchgears	High

Table 2.13 (Continued)

	Objective	Condition	Importance
(e)	• Assess synchronous machine rotor angle stability • Evaluate voltage variation and recovery • Assess system oscillations • Assess excitation protection responses	• N-1 contingency • Excitation system failure of generation unit i supplying a significant portion of the load • $i = 1, \ldots,$ number of main power generation units	Medium
(f)	• Assess synchronous machine rotor angle stability • Evaluate frequency variation and recovery • Assess system oscillations • Assess governor protection responses	• N-1 contingency • Speed governor system failure generation unit i supplying a significant portion of the load • $i = 1, \ldots,$ number of main power generation units	Medium
(g)	• Assess synchronous machine rotor angle stability • Evaluate frequency variation and recovery • Evaluate voltage variation and recovery • Assess system oscillations	• N-1 contingency • Trip of a large load (large motor, etc.) m • $m = 1, \ldots,$ number of large loads	High
(h)	• Assess synchronous machine rotor angle stability • Evaluate frequency variation and recovery • Evaluate voltage variation and recovery • Assess generator speed governor and excitation system responses • Review frequency and voltage protection performance • Assess the need for enhanced protections	• N-2 contingency • Trip of main power generation unit i and main power distribution equipment j in combination of sudden failure of critical switchgear l *(any two contingencies in combination)* • $i = 1, \ldots,$ number of main power generation units, $j = 1, \ldots,$ number of main power distribution equipment, $l = 1, \ldots,$ number of critical switchgears	Medium
(i)	• Assess synchronous machine rotor angle stability • Evaluate frequency variation and recovery • Evaluate voltage variation and recovery • Assess generator speed governor and excitation system responses • Review frequency and voltage protection performance • Assess the need for enhanced protections	• N-1-1 contingency • Trip of main power generation unit i and main power distribution equipment j in combination of sudden failure of critical switchgear l (any two contingencies in sequence) • For critical systems • $i = 1, \ldots,$ number of main power generation units, $j = 1, \ldots,$ number of main power distribution equipment, $l = 1, \ldots,$ number of critical switchgears	Medium

Table 2.14 Optimal power flow scenarios for optimization-based analysis.

	Objective	Condition	Importance

(a)	• Minimize system losses
	• Minimize energy costs
	• Minimize generation costs
	• Maximize voltage security
	• Maximize line/cable flow security
	• Minimize control efforts
	• etc.
		• Regular load flow scenarios
		• System operation constraints, i.e., benchmarked load flow cases, plus defined constraints including:
		  • Generation capacity
		  • Power import limit
		  • Line/cable thermal limits
		  • Bus voltage ranges
		  • Control variable ranges
		• Cost and price:
		  • Generation fuel cost
		  • Power import cost
		  • Power export price
		  • Operation cost
		  • Control cost
			Medium

Table 2.15 Short-circuit scenarios for optimization-based analysis.

Objective	Condition	Importance
• Minimize downtime • Minimize interruption area • Maximize critical load preservation • Minimize equipment damage • etc.	• Regular short-circuit scenarios • Protective device types and options • Protective device setting ranges	—

2.3.4.2 Short-Circuit and Protection Coordination Analysis

Short-circuit analysis scenarios in optimization stage should cover the benchmark study cases except those become not applicable due to equipment outages and system configuration changes and make efforts by adjusting protective device settings and relay coordination to minimize system downtime, minimize interrupting areas, and maximize preservation of critical loads. Table 2.15 shows corresponding analysis scenarios.

2.3.5 An Analysis Example

This example illustrates a complete system expansion design, study and analysis project of a major refinery plant [11]. Discussions are focused on study scenario selections for a number of analyses.

2.3.5.1 System Description

The plant is powered by two main power supplies from the local utility and in-plant generators. Incoming 66 kV lines from the utility are connected to the refinery through two 66 to 22 kV step-down transformers. After the expansion, the refinery has one 12.5 MVA existing steam turbine generator (TA), two 56.562 MVA new gas turbine generators (GTG1 and GTG2), and one 2.438 MVA new steam turbine alternator (STA). Electrical loads are connected at 3.3 kV and 420 V through the sub-distribution power system. The refinery electrical system consists of 14 substations. Four of the substations are new as part of the expansion and are included in the scope of this study project. They are:

1) 22 kV main substation (MAIN)
2) 3.3 kV co-gen substation (COGEN)
3) 3.3 kV new substation 1 (SUB1)
4) 3.3 kV new substation 2 (SUB2)

Substation bus configurations are all in main-tie-main with tie breakers normally open. When there is a need to close tie breakers, the tie breakers in the 3.3 kV system should be closed before those in the low voltage switchgears. Due to the direct installation of new GTG generators on 22 kV buses, the main power distribution switchgear was replaced with a new gas-insulated switchgear (GIS) system in 22 kV main substation to eliminate fault exposures. In addition, there was a new process train being incorporated into the existing system with a new substation added, and modification of the existing power distribution system was required. On the other hand, increased short circuit current duties at 22 kV buses suggest a CLR installed at the bus tie position of the 22 kV GIS to reduce the contribution of the fault current from the other bus bar. There are also two other CLRs connected between the GIS and downstream substations. The primary objective of the study documented in this paper was to determine the impact of the proposed changes to the overall system, including new in-plant generators, new process train, new co-gen plant, new GIS and new CLR. The studies were performed in each part of the project phase. The system conceptual single-line diagram is shown in Figure 2.13.

2.3.5.2 Operation Modes and Constraints

The anticipated refinery power demand is 77 MW. Some future reserve demand may be added to take the total refinery load to 80 MW. Based on equipment rating and design specification, the total output from 2 GTGs is 57 MW. Combined power generation from TA and STA is around 8 MW. The rest of the power will be imported from the utility. The system must meet certain operation mode requirements and constraints from the utility which are explained below.

Table 2.16 Load demands and power supplies under normal operation mode.

Load demand		Power supply	
Description	**MW**	**Description**	**MW**
Power demand	77	2 GTGs	57
		Existing steam turbine generator TA	7
		New steam turbine alternator STA	1
		Import from the utility	12

1) *Operation modes.* The system was designed to operate in two primary operation modes according to the refinery "Operation Philosophy" at the time of the studies:
   - *Normal operation.* In this mode, the refinery is in parallel operation with the utility. Typical power balance in normal operation mode is shown in Table 2.16.
   - *Island operation.* In this mode, the refinery is separated from the utility. The island operation mode can be initiated either as a planned event or a sudden event. In the planned event case, the island operation is anticipated, the refinery is prepared for reduced operation prior to starting the island operation. In the sudden event case, a circuit failure occurs at the utility, and the refinery power generation system needs to be changed to island operation mode. The 22 kV MAIN substation electrical operator will reduce the refinery power demand in accordance with a pre-determined procedure. Power balance during island operation mode is shown in Table 2.17.
2) *Operation constraints.* In addition to internal operation modes, the utility requires to: (1) limit fault current contribution from the plant to the grid under a certain level (around 5.0 kA), (2) achieve a higher power factor (nearly unity) at the PCC interface with the grid.

Table 2.17 Planned islanded operation mode.

Load demand		Power supply	
Description	**MW**	**Description**	**MW**
Power demand	65	2 GTGs	57
		Existing steam turbine generator TA	7
		New steam turbine alternator STA	1

2.3.5.3 Studies Scenario Selections

To study the new system, a system computer model for the entire system needed to be completed. The existing system computer model was available from the EPC. This model was expanded to include the new system and components, initially done by the EPC in the system modeling stage. However, the final system model was completed and validated through this project. A considerable number of man-hours was spent in the process for data verification. The scope of work for this project did not include data collection but it is important to point it out.

The next step was to decide what types of study were needed. Per industry standards [1, 2], electrical power studies normally include load flow, short circuit, motor starting, protection coordination, and harmonics if there are excessive harmonic sources and exist power quality concerns, plus transient stability if there are on-site generators. Arc flash is also required by National Fire Protection Association (NFPA) requirement [12] using the IEEE recommended method [13]. The scope of this project includes all studies listed above except harmonics and arc flash.

Finally, study cases were selected based on designed normal, abnormal operation and maintenance modes, major N-1 contingencies, and industry standards. Each type of the studies had some system configuration variations to reflect study type dependent conditions. Cases for each type of study and the rationales of selecting these cases are described in this section.

1) *Network topology*. The network topology includes a normal operation configuration and several contingency configurations, which are shown in Figure 2.14 through Figure 2.17.
   - Figure 2.14 represents the system operating in the normal operation mode. Both incoming 66 kV lines from two utility substations are connected and all generators are running at normal power generation level. The 22 kV buses are connected through the closed tie breaker and the CLR.
   - Figure 2.15 represents an operating condition under a contingency when there is only one utility supply (one of the utility lines is out of service, or one of the utility main transformers is out of service) and one GTG on the same side of the energized network is running. All other generators are offline. All the downstream buses are operated with a single input and the tie breakers are closed.
   - Figure 2.16 represents another operating condition under contingency when there is only one utility supply and one GTG on the other side of the network is running. The 22 kV buses are connected through the closed tie breaker and the CLR. All other generators are offline. All the downstream buses are operated with a single input and the tie breakers are closed.
   - Figure 2.17 represents the planned islanded operation mode when both utility supplies are lost. The 22 kV buses are directly connected through the

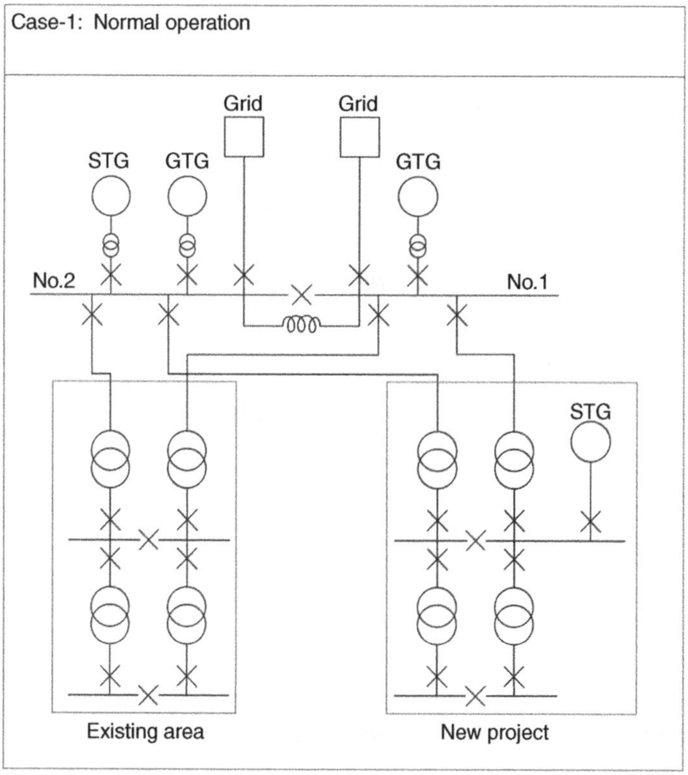

Figure 2.14 Normal operation. Source: Dai [11]/with permission of IEEE [4].

closed tie breakers but without the CLR. All generators are running. All the downstream buses operate in the normal tie breakers open status.

For each type of study, there are possible variations from the four main configurations illustrated in Figures 3.14–3.17 to meet specific study objectives.

2) *Load flow analysis*. The load flow study includes four study cases as shown in Table 2.18. These cases reflect the four configurations defined in Figures 3.14–3.17.

For the load flow study, normal loading was considered. Generator voltage settings and transformer taps were adjustable. The objectives are to check and evaluate system operating conditions at steady state for equipment loading, voltage regulation, generator capacity, and in particular for this project the CLR loading, and voltage drop.

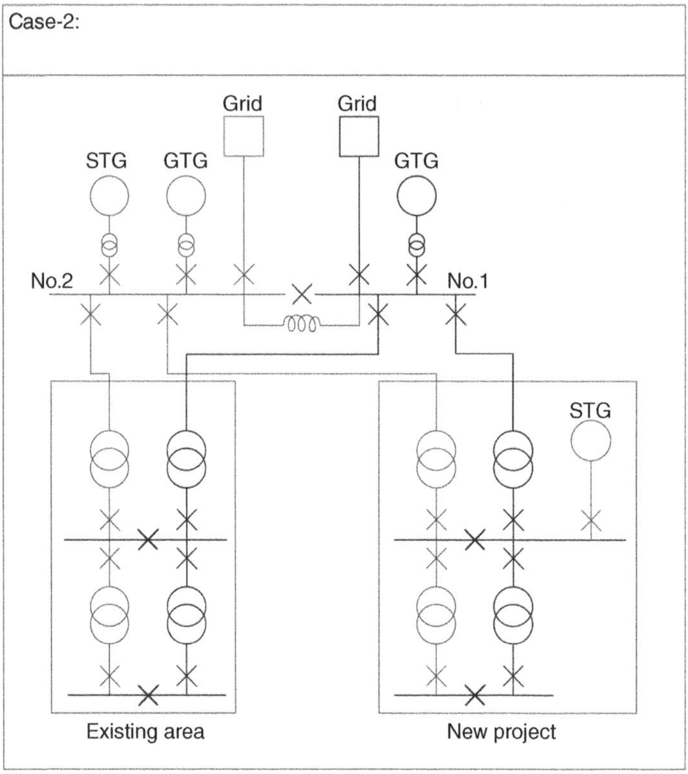

Figure 2.15 One utility supply. Source: Dai [11]/with permission of IEEE [4].

The load flow study results were used to assist in the following design and adjustment under all configurations including normal and contingency scenarios:

- Sizing equipment (new cable, transformer, and CLR)
- Determining equipment settings (generator AVR, transformer tap)
- Identifying low voltage situations (system wide)
- Identifying overloading conditions (generator, cable, transformer)
- Checking PCC constraints
- Mitigating any violations by coordinating between adjusting various control settings and resizing equipment

3) *Short circuit study*. Six study cases were performed for short circuit study as shown in Table 2.19. These cases were designed to reflect the highest fault current duties under various configurations.

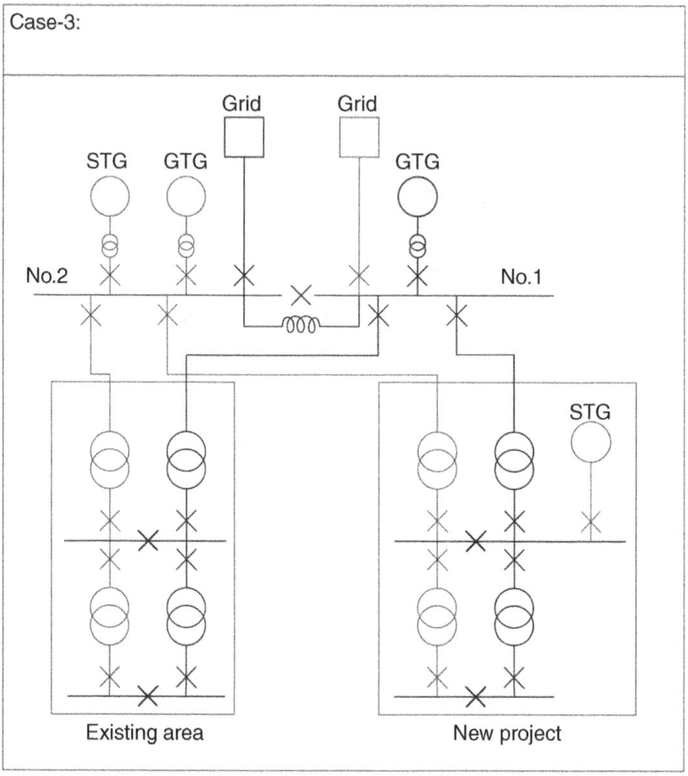

Figure 2.16 One utility with CLR. Source: Dai [11]/with permission of IEEE.

Three- and single-phase faults were simulated under the given configurations and scenarios at pertinent locations throughout the area of the project scope. Fault current contributions from all motors were included. The objectives are to determine the maximum short circuit currents at each bus. The short circuit study results were used to:

- Evaluating MV breaker making and interrupting capabilities
- Resizing MV breakers (when necessary)
- Evaluating LV breaker interrupting capabilities
- Resizing LV breaker (when necessary)
- Evaluating bus bar making capabilities
- Evaluating CLR current limiting effect

4) *Protection and coordination study*. The system coordination begins at the 66 kV MV relays and continues downstream through to feeder breakers at each panel

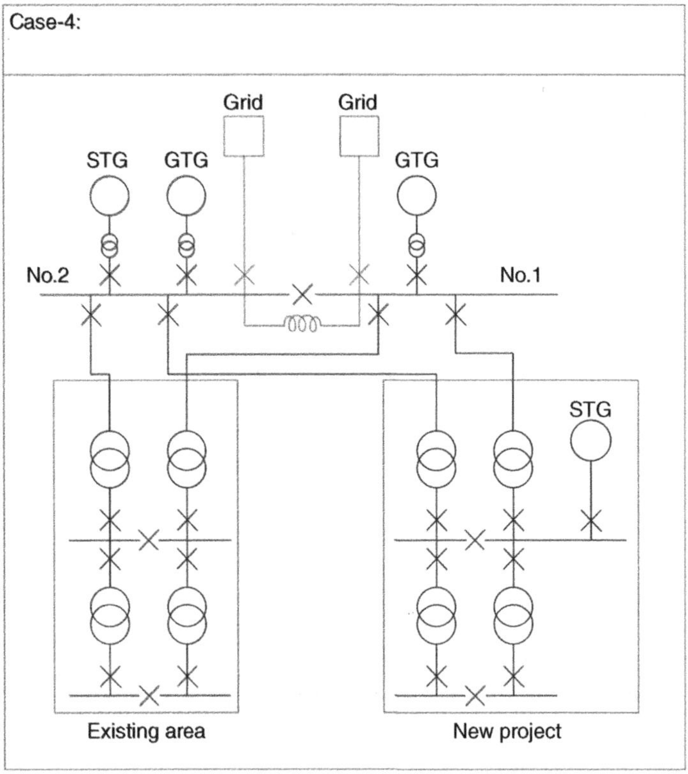

Figure 2.17 Planned island operation. Source: Dai [11]/with permission of IEEE.

Table 2.18 Load flow analysis cases.

Case no.	Description
Case 1	Normal operation: all bus ties open, 22 kV CLR in place
Case 2	One utility supply and one GTG on the same side of the network, all bus ties closed, one power train out
Case 3	One utility supply and one GTG on another side of the network, all bus ties closed, one power train out, 22 kV CLR in place
Case 4	Island operation with both utility supplies out, 22 kV buses tied directly

Table 2.19 Summary of short circuit analysis cases.

Case no.	Description
Case 1	Normal operation: all ties open except 22 kV CLR in place
Case 2	Normal operation: all ties open except 22 kV tied directly
Case 3	Only one utility supply, all ties open except 22 kV tied directly
Case 4	One GTG out of service, all ties open except 22 kV tied directly
Case 5-A	One utility supply and one GTG out of service, one power train out of service, all ties closed, no CLR
Case 5-B	Both utility supplies out of service, one power train out of service, all ties closed, no CLR

Table 2.20 Protection zones and TCC curves.

Zone (from – to)	No. of TCC curves
GTG – GIS	4
Existing TA – GIS	2
Utility supply – GIS	3
GIS supply – SUB1	4
Existing 3.3 kV SWG – COGEN 3.3 kV SWG	4
COGEN 3.3 kV SWG – LV SWG	4
COGEN 3. 3 kV SWG – MV Motor	4
COGEN typical LV motor	4
SUB1 3.3 kV SWG – LV SWG	4
SUB1 3.3 kV SWG – MV motor	8
SUB1 0.42 kV SWG – LV motor	2
New STA – Existing 3.3 kV SWG	2
COGEN 0.42 kV SWG – LV motor	10
Total TCC Curves	55

board in the new expansion system. A total of 55 TCC curves were created and validated, as illustrated in Table 2.20.

Three- and single-phase fault duties calculated from the short circuit study were used in the coordination studies. The general objectives of the protection and coordination study are to review the system's overcurrent protection and coordination functions and provide suggestions for improvement. One special objective for this project is to provide the best coordination possible with the protective devices in the existing system. Areas where breaker tripping curves overlapped indicated areas of possible non-selective breaker operation. Where possible, efforts were made to reduce nonselective breaker operation to maintain adequate system protection. In some cases, because of device limitations, little can be done to improve device selectivity, and these cases were highlighted in the report. Examples of such device limitations are fixed operating characteristic of fuses and built-in instantaneous trip units.

5) *Motor starting study*. Four study cases were defined for motor starting study as shown in Table 2.21. These cases follow the four configurations defined in Figure 2.14 through Figure 2.17. Each study case includes multiple subcases to study starting the largest motor on each bus.

Motor starting studies were performed using dynamic starting simulation with motor circuit model, inertia, and load curve. The objectives are to evaluate the voltage reduction and other impacts of starting motors under given configurations and determine if the system has sufficient capacity to start the motors in the area of the scope of work while maintaining bus voltage levels for locations per the voltage criteria below:

Motor terminal voltage ≥ 80%
Other bus voltage ≥ 85%

6) *Transient stability analysis*. Six study cases were selected for transient stability study as shown in Table 2.22. These cases include scenarios of faults on generator terminal buses, loss of utility supplies, fault isolation under various configurations as defined in Figures 3.14–3.17.

Generator parameters, excitation and governor models and parameters are listed in [11]. The objectives are to assess generator rotor angle stability, system frequency and voltage variations, generator governor and excitation systems

Table 2.21 Motor starting analysis cases.

Case no.	Description
Case 1	Normal operation: all bus ties open, 22 kV CLR in place
Case 2	One utility supply and one GTG on the same side of the network, all bus ties closed, one power train out
Case 3	One utility supply and one GTG on another side of the network, all bus ties closed, one power train out, 22 kV CLR in place
Case 4	Island operation with both utility supplies out, 22 kV buses tied directly

Table 2.22 Transient stability analysis cases.

Case no.	Description
Case 1	Normal operation mode: three-phase fault at a 22 kV feeder then cleared by line differential
Case 2	Normal operation mode: sudden loss of both utility supplies
Case 3	One GTG in maintenance mode and 22 kV buses directly tied: sudden loss of one utility supply
Case 4	Normal operation mode: three-phase fault near STG followed by fault clearing
Case 5	Normal operation mode: three-phase fault at a utility supply followed by separating the faulted utility supply
Case 6	Planned islanded mode: three-phase fault at a 22 kV feeder then tripped by line differential

controllability, and impact to operating loads under various transient conditions. Transient stability study results were used to assist in:

- Determining or validating fault clearing protection relay settings
- Assessing system frequency variation and determining needed remedies
- Assessing voltage variation and determining needed protections
- Evaluating speed governor and excitation control capabilities and determining needed protections
- Studying system island operation capabilities and determining needed load shedding protections

7) *Analyses during different stages of project.* An after-FEED detailed study normally covers several stages of the project, and the system database will evolve from the initial as-design to the final as-build along with the project execution. At each stage, the system database may change in accordance with the progress in equipment procurement and updated vendor information. It is important to maintain the latest computer model and database during the entire course of the project. Sometimes the study cases may also need to be modified to reflect design, equipment, and other condition changes. System studies will be performed at each stage based on the concurrent database, and the new study reports shall be submitted to the EPC company and the plant owners promptly. This allows continuous validation of the system design, operation, and construction. It is, therefore, very vital to schedule resources to follow the stages of a project for performing system studies and preparing reports. In this project, a total of three sets of complete system study were completed over one year period for stages of FR, FD and FC respectively. Several system configurations,

operating philosophies and equipment data went through update process during different stages, including:

- STA nameplate data
- GTG governor and exciter system model parameters
- Transformer data
- Future load amount and connecting locations, etc.

2.4 Data and Results

To perform I&C power system study and analysis, different data sets are required. Some data sets are about description of the electrical system structure and configurations, equipment specifications and characteristics, load profiles, power source capabilities, protection curves and settings, control system model and parameters, and any other data associated with the electrical components in the system that will participate in system operation and affect system responses. Some other data are related to study scenarios and conditions which are case by case to define specific system initial operation conditions and sometimes events to drive time-domain simulations. Some studies will involve additional data which are either mandatory, or model related optional, or calculation algorithm required, or standard provided. Finally output data are generated that provides calculation results from computer modeling and simulation. Data sets can vary depending on the specific analysis being conducted, but there are some common types of data by all studies. The mentioned data sets are explained in the following sections.

2.4.1 Input Data

Input data should include technical specifications for equipment ratings, impedance values, and other relevant parameters. Following input data should be collected and ready at the stage of system modeling before conducting studies and analysis.

Generic load data. Generic loads represent Heating, Ventilation, and Air Conditioning (HVAC) load, lighting, computer, general IT load or other wide range non-motor loads. These loads, together with motor loads, are also called load schedule or load profile which provide information on all connected loads to the system. Table 2.23 lists typical generic load data and descriptions as specified in a load panel as well as those required for simulation analysis. Figure 2.18 is a sample load panel datasheet.

Motor data. Motor data include motor ratings, starting parameters, and motor dynamic models. Tables 2.24–2.26 list typical motor nameplate data, single-cage machine and double-cage circuit models and descriptions. Figure 2.19 is a sample motor nameplate. Motor dynamics can also be expressed in characteristic

Table 2.23 Load data.

	Data piece	Description	Note
(a)	Item no.	Panel name	
(b)	Type	Load type (constant impedance, constant power, constant current, or others)	
(c)	VOLTS	Rated voltage	
(d)	LOAD DESCRIPTION A	Description of phase A load	
(e)	LOAD DESCRIPTION B	Description of phase B load	
(f)	LOAD DESCRIPTION C	Description of phase C load	
(g)	kVA A	Rated power of phase A load	
(h)	kVA B	Rated power of phase B load	
(i)	kVA C	Rated power of phase C load	
(j)	AMP A	Rated current of phase A load	
(k)	AMP B	Rated current of phase B load	
(l)	AMP C	Rated current of phase C load	
(m)	COND SIZE A	Conduit size of phase A	
(n)	COND SIZE B	Conduit size of phase B	
(o)	COND SIZE C	Conduit size of phase C	
(p)	WIRE SIZE A	Conductor wire size of phase A	
(q)	WIRE SIZE B	Conductor wire size of phase B	
(r)	WIRE SIZE C	Conductor wire size of phase C	
(s)	PF % A	Rated power factor of phase A load	
(t)	PF % B	Rated power factor of phase B load	
(u)	PF % C	Rated power factor of phase C load	
(v)	CONNECT	Load phase connection (Delta, Y)	
(w)	GROUND	Load grounding (Grounded, open)	
(x)	Category	Loading category i name ($i = 1, \ldots,$ maximum loading category)	
(y)	% Loading	Loading percentage on the base of the rated power for loading category i ($i = 1, \ldots,$ maximum loading category)	

A Ø:	18.59	kVA	EXISTING PANEL S3A	3	Ø	4	W
B Ø:	17.77	kVA				120/208	VOLT
C Ø:	19.40	kVA	3P-200A M.C.B		225	AMP	BUS
TOTAL	55.76	kVA	155 AMPS	BREAKER INTERRUPT RATING:		10,000	A.I.C.

COND SIZE	WIRE SIZE	LOAD DESCRIPTION	kVA A Ø	kVA B Ø	kVA C Ø	C/B	CKT NO.	CKT NO.	C/B	kVA A Ø	kVA B Ø	kVA C Ø	LOAD DESCRIPTION	WIRE SIZE	COND SIZE
		(EX)LTG. PARTY	1.60			1P-20	1	2	IP-20	1.50			(EX)HAND DRYER, WOMAN		
		(EX)LTG. VERANDA		0.80		1P-20	3	4	I		1.50		(EX)HAND DRYER, WOMAN		
		(EX)LTG. VERANDA			0.90	1P-20	5	6	I			1.50	(EX)HAND DRYER, MEN		
		(EX)LTG. VERANDA	0.80			1P-20	7	8	I	1.50			(EX)HAND DRYER, MEN		
		(EX)LTG. VERANDA		0.90		1P-20	9	10	I		0.36		(EX)RECEPTS, WOMEN		
		(EX)LTG. WOMEN/MEN			1.30	1P-20	11	12	I			0.36	(EX)RECEPTS, MEN		
		(EX)LTG. STAIR	0.50			1P-20	13	14	I	0.36			(EX)RECEPTS, SND EQUIP		
		(EX)PLUMB FIXT SENSORS		0.20		1P-20	15	16	I		0.54		(EX)RECEPTS, PARTY		
1/2"	2#12	MOTORIZED SCRNS NW			1.36	1P-20	17	18	I			0.54	(EX)RECEPTS, PARTY		
I	I	MOTORIZED SCRNS N	1.11			1P-20	19	20	I	0.90			(EX)RECEPTS, VERANDA		
I	I	MOTORIZED SCRNS NE		1.36		1P-20	21	22	I		1.08		(EX)RECEPTS, VERANDA		
V	V	MOTORIZED SCRNS SW			1.36	1P-20	23	24	I			0.72	(EX)RECEPTS, VERANDA		
		(EX)AHU-1 HEAT	3.30			3P-40	25	26	I	0.72			(EX)RECEPTS, VERANDA		
		I		3.30			27	28	I		0.72		(EX)RECEPTS, VERANDA		
		V			3.30		29	30	I			0.80	(EX)DRINKING FOUNTAIN		
		(EX)AHU-3 HEAT	3.30			3P-40	31	32	I	0.40			(EX)REC/DMP, VER/PRTY		
		I		3.30			33	34	I		1.11		MOTORIZED SCRNS S	2#12	1/2"
		V			3.30		35	36	V			1.36	MOTORIZED SCRNS SE	V	V
		(EX)AHU-3 FAN	1.30			3P-20	37	38	3P-20	1.30			(EX)AHU-1 FAN		
		I		1.30			39	40			1.30		I		
		V			1.30		41	42				1.30	V		
F.T.L.															

Figure 2.18 Sample load panel nameplate or datasheet. Source: Adapted from https://www.bing.com/images/search?view=detailV2&ccid=yUq1xdyw&id=D05D0A2259ABD44EB3DDE6601F620AA3A60B7342&thid=OIP.yUq1xdywSKR_6VYTbCsZrwHaD_&mediaurl=https%3a%2f%2fforums.autodesk.com%2ft5%2fimage%2fserverpage%2fimage-id%2f332645i1DAB9D35422A2FEA%2fimage-size%2flarge%3fv%3d1.0%26px%3d999&cdnurl=https%3a%2f%2fth.bing.com%2fth%2fid%2fR.c94ab5c5dcb048a47fe956136c2b19af%3frik%3dQnMLpqMKYh9g5g%26pid%3dImgRaw%26r%3d0&exph=539&expw=999&q=load+switchboard+nameplate+picture&simid=607996473091312232&FORM=IRPRST&ck=2B99DC1507512D036B23DA699323DD42&selectedIndex=59&itb=0&ajaxhist=0&ajaxserp=0/Last accessed on 05 Sep, 2024.

Table 2.24 Motor nameplate data.

	Data piece	Description	Note
(a)	Item no.	Equipment name or ID	
(b)	PHASE	Number of phases	
(c)	HP	Horsepower	
(d)	FREQ	Frequency	
(e)	RPM	Speed	
(f)	VOLTS	Rated voltage	
(g)	F.L. AMPS	Full load current	
(h)	S.F. AMPS	Service factor current	
(i)	S.F.	Service factor	
(j)	DESIGN	NEMA design type	
(k)	CODE	Design code	
(l)	NEMA NOM. EFF.	NEMA nominal efficiency	
(m)	NEMA MIN. EFF.	NEMA minimum efficiency	
(n)	NOM. P.F.	Nominal power factor	
(o)	MAX. KVAR	Maximum kVAR	
(p)	CONNECT	Motor phase connection (Delta, Y)	
(q)	GROUND	Motor grounding (Grounded, open)	
(r)	Category	Motor loading category i name ($i = 1, \ldots,$ maximum motor loading category)	
(s)	% Loading	Motor loading percentage on the base of the rated power for motor loading category i ($i = 1, \ldots,$ maximum loading category)	
(t)	T_{st}	Motor starting torque	
(u)	PF_{st}	Motor starting power factor	
(v)	I_{st}	Motor starting in-rush current	
(w)	T_b	Motor break in torque	
(x)	Motor model	Motor circuit model or torque-slip curve	
(y)	H	Motor inertia constant	
(z)	D	Motor damping factor	

Table 2.25 Single-cage motor circuit model data.

	Data piece	Description	Note
(a)	R_S	Motor stator resistance	% on machine base
(b)	X_s	Motor stator reactance	% on machine base
(c)	X_m	Motor magnetizing reactance	% on machine base
(d)	R_C	Motor core loss resistance	% on machine base
(e)	$R_{r,fl}$	Motor rotor resistance at full load	% on machine base
(f)	$R_{r,lr}$	Motor rotor resistance at locked-rotor	% on machine base
(g)	$X_{r,fl}$	Motor rotor reactance at full load	% on machine base
(h)	$X_{r.lr}$	Motor rotor reactance at locked-rotor	% on machine base

Table 2.26 Double-cage motor circuit model data.

	Data piece	Description	Note
(a)	R_S	Motor stator resistance	% on machine base
(b)	X_s	Motor stator reactance	% on machine base
(c)	X_m	Motor magnetizing reactance	% on machine base
(d)	R_C	Motor core loss resistance	% on machine base
(e)	R_{r1}	Motor rotor resistance for the first rotor circuit	% on machine base
(f)	R_{r2}	Motor rotor resistance for the second rotor circuit	% on machine base
(g)	X_{r1}	Motor rotor reactance for the first rotor circuit	% on machine base
(h)	X_{r2}	Motor rotor reactance for the second rotor circuit	% on machine base

curves which include motor torque versus speed, power factor versus speed and efficiency versus speed. Table 2.27 lists typical motor characteristic curve data in table format and descriptions.

Motor load data. Motor load data basically is load torque versus speed. Figure 2.20 is a sample fun-type load torque vs speed characteristic curve.

Utility data. Utility data do not have standard format or nameplate. Essentially, they should include the information for load flow, short-circuit, and harmonic analysis. Table 2.28 lists the essential utility data.

Generator data. Machine rating, machine model, governor and prime mover model, and excitation and AVR model data for synchronous generator are shown in Tables 2.29 and 2.30. Figure 2.21 is a sample synchronous generator nameplate.

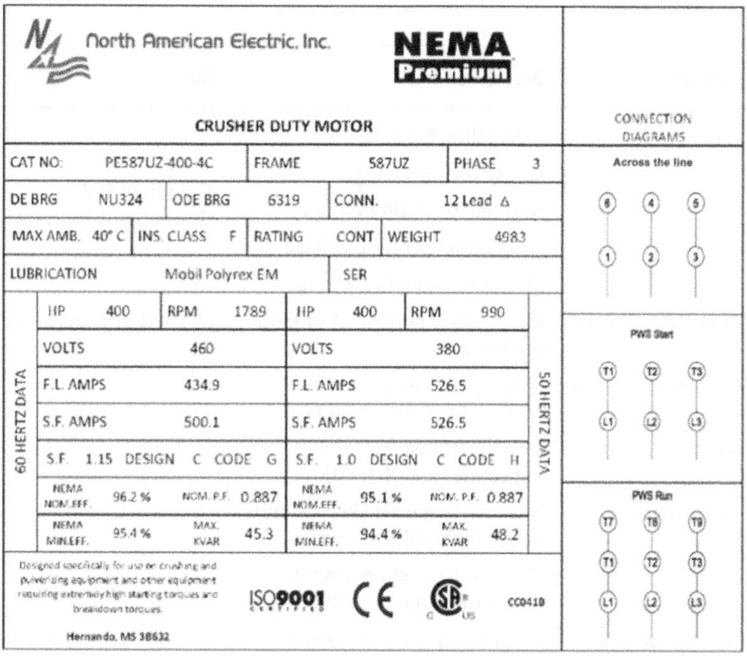

Figure 2.19 Sample motor nameplate. Source: North American Electric, Inc.

Table 2.27 Motor characteristic curve model data.

	Data piece	Description	Note
(a)	{Torque *i*, Speed *i*}	Data series of % load torque at % speed at point *i* ($i = 1, \ldots,$ maximum number of points)	Torque and speed based on the motor rating
(b)	{Power Factor *i*, Speed *i*}	Data series of % load power factor at % speed at point *i* ($i = 1, \ldots,$ maximum number of points)	Speed based on the motor rating
(c)	{Efficiency *i*, Speed *i*}	Data series of % load efficiency at % speed at point *i* ($i = 1, \ldots,$ maximum number of points)	Speed based on the motor rating

Table 2.28 Utility data.

	Data piece	Description	Note
(a)	Item no.	Utility/line name	
(b)	kV	Utility voltage rating	Load flow and short-circuit analyses
(c)	V_{nom}	Nominal operating voltage	
(d)	V_{max}	Maximum operating voltage	
(e)	V_{min}	Minimum operating voltage	
(f)	MVA or Amp 3-phase	Three-phase fault MVA or fault current at PCC	Short-circuit analysis
(g)	MVA or Amp 1-phase	Single-phase-to-ground fault MVA or fault current at PCC	
(h)	Z_1 and X_1/R_1	Positive sequence short-circuits impedance and X/R ratio	
(i)	Z_2 and X_2/R_2	Negative sequence short-circuits impedance and X/R ratio	
(j)	Z_0 and X_0/R_0	Zero sequence short-circuit impedance and X/R ratio	
(k)	$Z_1(h)$ and $X_1(h)/R_1$	Positive sequency frequency dependent impedance and X/R ratio at harmonic order h $(h = 1, 2, \ldots, n)$	Harmonic analysis
(l)	$Z_2(h)$ and $X_2(h)/R_2$	Negative sequency frequency dependent impedance and X/R ratio at harmonic order h $(h = 1, 2, \ldots, n)$	
(m)	$Z_0(h)$ and $X_0(h)/R_0$	Zero sequency frequency dependent impedance and X/R ratio at harmonic order h $(h = 1, 2, \ldots, n)$	
(n)	$V_{mag}(h)$	Harmonic voltage spectrum given by voltage magnitude at harmonic order h $(h = 1, 2, \ldots, n)$	
(o)	$V_{ang}(h)$	Harmonic voltage spectrum given by voltage angle at harmonic order h $(h = 1, 2, \ldots, n)$	
(p)	$I_{mag}(h)$	Harmonic current spectrum given by voltage magnitude at harmonic order h $(h = 1, 2, \ldots, n)$	
(q)	$I_{ang}(h)$	Harmonic current spectrum given by voltage angle at harmonic order h $(h = 1, 2, \ldots, n)$	

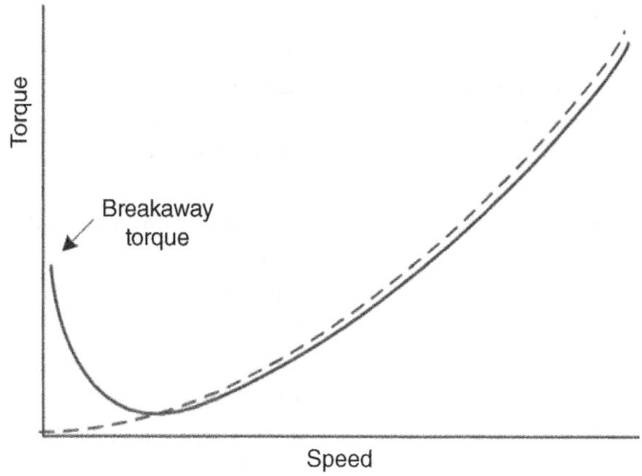

Figure 2.20 Sample motor load characteristic curve. Source: Mutize, Cephas & Wang, Rong-Jie. (2013).

Table 2.29 Generator rating data.

	Data piece	Description	Unit
(a)	Item no.	Equipment ID	
(b)	Manufacturer	Manufacturer name	
(c)	Standard	Standard type (e.g., ANSI, IEC, …)	
(d)	Output kVA/MVA	Rated apparent power output in kVA or MVA	
(e)	Voltage	Rated voltage	
(f)	Current	Rated current	
(g)	Frequency	Rated frequency	
(h)	Poles	Number of poles	
(i)	Frequency	Rated frequency	
(j)	Phase	Number of phases	
(k)	Ex. Voltage	Excitation voltage	
(l)	Ex. Current	Excitation current	
(m)	Power Factor	Rated power factor	
(n)	Efficiency	Rated efficiency	

Generator excitation and AVR system, and turbine and speed governor system data are described in Tables 2.31 and 2.32 where details about transfer function block diagram of the two systems will be discussed in Section 3.2.1.

Transformer data. Transform data are shown in Table 2.33. Figure 2.22 is a sample transformer nameplate.

Table 2.30 Generator machine data.

	Data piece	Description	Unit
(a)	$X_{d,sat}, X_{q,sat}$	Synchronous saturated d-axis and q-axis reactance	p.u.
(b)	$X_{d,unsat}, X_{q,unsat}$	Synchronous unsaturated d-axis and q-axis reactance	p.u.
(c)	$X'_{d,sat}, X'_{q,sat}$	Transient saturated d-axis and q-axis reactance	p.u.
(d)	$X'_{d,unsat}, X'_{q,unsat}$	Transient unsaturated d-axis and q-axis reactance	p.u.
(e)	$X''_{d,sat}, X''_{q,sat}$	Subtransient saturated d-axis and q-axis reactance	p.u.
(f)	$X''_{d,unsat}, X''_{q,unsat}$	Subtransient unsaturated d-axis and q-axis reactance	p.u.
(g)	X_2	Negative sequence reactance	p.u.
(h)	X_0	Zero sequence reactance	p.u.
(i)	X_L	Leakage reactance	p.u.
(j)	$T_{d0,sat}', T_{q0,sat}'$	Transient open circuit saturated d-axis and q-axis time constant	sec
(k)	$T_{d0,unsat}', T_{q0,unsat}'$	Transient open circuit unsaturated d-axis and q-axis time constant	sec
(l)	$T_{d0,sat}'', T_{q0,sat}''$	Subtransient open circuit saturated *d*-axis and *q*-axis time constant	sec
(m)	$T_{d0,unsat}'', T_{q0,unsat}''$	Subtransient open circuit unsaturated d-axis and q-axis time constant	sec
(n)	$T_{d,sat}'', T_{q,sat}''$	Subtransient short-circuit saturated d-axis and q-axis time constant	sec
(o)	$T_{d,unsat}'', T_{q,unsat}''$	Subtransient short-circuit unsaturated d-axis and q-axis time constant	sec
(p)	T_a	Armature DC component time constant Three phase short-circuit	sec.
(q)	R_a	Armature winding DC resistance per phase	Ohm
(r)	X''/R_a	Armature X/R ratio	p.u.
(y)	J	Rotor inertia (generator only)	kg m^2
(z)	H	Turbine – generator combined inertia constant	kW-sec/kVA
(aa)	D	Shaft mechanical damping	% MW/Hz
(bb)	Δ/Y	Generator winding connection	Delta or Wye
(cc)	Grounding type	Generator winding grounding method	
(dd)	Grounding impedance	Generator winding grounding impedance for *Y* connection	p.u.

Synchronous Generator			
Type	GDH6529S-14	INSUL. CLASS	F
Output	1500/17250 kVA	FREQUENCY	
Voltage	13800 V	POLES	60 Hz
Current	628/722 A	PHASES	28
Ex. Voltage	230 V	RATING	3
Ex. Current	590/645 A	TEMP. RISE	CONT.
Power Factor	0.85	AMB. TEMP.	SR 75/90 RT 80/100°C
Speed	257 rpm		40 °C
Standard	ANSI C50. 12-1982	SERIAL NO.	KC69008L1
	MANUFACTURED	1991	

Figure 2.21 Sample generator nameplate. Source: https://www.google.com/search?q=+synchronous+generator+nameplate+pictures&sca_esv=7f561bb85968220c&biw=1248&bih=639&ei=6y4GZrGnOffJkPIPi9G76As&ved=0ahUKEwjxyN7LvZiFAxX3JEQIHYvoDr0Q4dUDCBA&uact=&oq=+synchronous+generator+nameplate+pictures&gs_lp=Egxnd3Mtd2l6LXNlcnAiKSBzeW5jaHJvbm91cyBnZW5lcmF0b3IgbmFtZXBsYXRlIHBpY3R1cmVzMggQABiABBiiBDIIEAAYgAQYogQyCBAAGIAEGKIEMggQABiABBiiBEjeAlAAWABwAHgAkAEAmAFWoAFWqgEBMbgBA8gBAPgBAZgCAaACWJgDAJIHATGgB8QC&sclient=gws-wiz-serp#vhid=wZGD4BkH7XC3BM&vssid=l/ with permission of Learneo, Inc./Last accessed on 05 Sep 2024.

Table 2.31 Generator excitation and AVR system data.

	Data piece	Description	Unit
(a)	Type	Excitation and AVR system type or model name	
(b)	Control function	Excitation and AVR system transfer function block diagram	

Table 2.32 Generator turbine and speed governor system data.

	Data piece	Description	Unit
(a)	Turbine/engine type	Turbine/engine type or model name	
(b)	Turbine/engine model	Turbine/engine model transfer function block diagram	
(c)	Governor type	Governor type or model name	
(d)	Governor control	Speed governor control transfer function block diagram	

Table 2.33 Transformer data.

	Data piece	Description	Note
(a)	Item no.	Equipment name or ID	
(b)	kV-A	Rated power	
(c)	HV V	Rated primary side voltage	
(d)	LV V	Rated secondary side voltage	
(e)	HV A	Rated primary side current	
(f)	LV A	Rated secondary side current	
(g)	BIL	Basic insulation level	
(h)	Connection	Connection type (Dela, Y)	
(i)	Frequency	Rated frequency	
(j)	Grounding	Grounding type (grounded, non-grounded)	
(k)	Grounding X	Grounding impedance	
(l)	X_{sc}	Short-circuit impedance	
(m)	Losses	Full load losses	
(n)	No-load losses	No-load losses	
(o)	Temperature	Nominal ambient temperature	
(p)	Tap/HV V	Tap position on the primary side	

Table 2.34 Reactor data.

	Data piece	Description	Note
(a)	Item no.	Equipment name or ID	
(b)	L or Z%	Inductance or impedance in percent	
(c)	V	Rated voltage	
(d)	V_{ins}	Insulation voltage rating	
(e)	I_{nom}	Nominal current	
(f)	I_{sat}	Saturation current rating	
(g)	f	Rated frequency	
(h)	T	Ambient temperature	

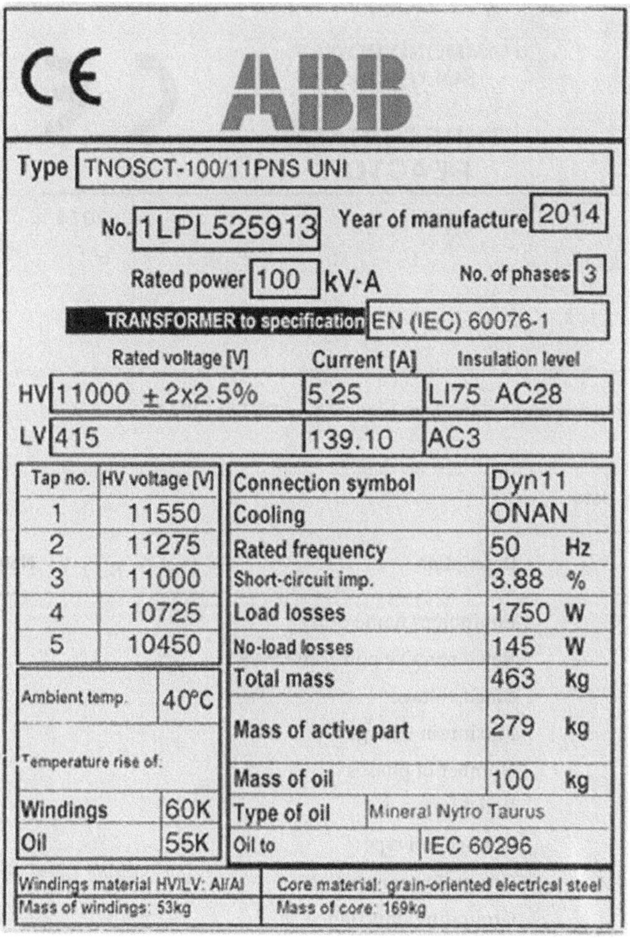

Figure 2.22 Sample transformer nameplate. Source: ABB Ltd.

Reactor data. Reactor data are shown in Table 2.34. Figure 2.23 is a sample reactor nameplate.

Capacitor data. Capacitor data are shown in Table 2.35. Figure 2.24 is a sample capacitor nameplate. Capacitor data may also include the sensor information

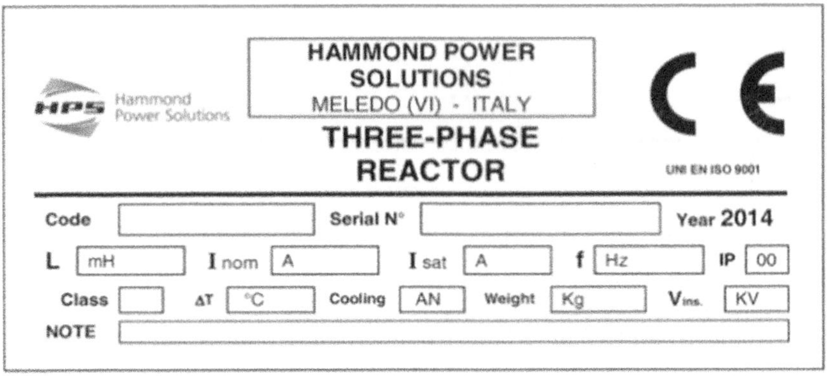

Figure 2.23 Sample reactor nameplate. Last accessed on 05 Sep, 2024.

Table 2.35 Capacitor data.

	Data piece	Description	Note
(a)	Item no.	Equipment name of ID	
(b)	KVAR	Rated reactive power in kVAR	
(c)	VOLTS	Rated voltage	
(d)	Vmax	Maximum voltage	
(e)	PH	Number of phases	
(f)	LOSSES	Rated losses	
(g)	CONN	Connection type	
(h)	GRND	Grounding connection	
(i)	X_{grd}	Grounding impedance	
(j)	{Position i, kVAR i}	Data series of fixed switching position i and capacitor kVAR ($i = 1, \ldots,$ maximum number of points)	
(k)	POS	Fixed switching position	

of what is the control signal and where it is coming from which are not listed.

Breaker data. Breaker data is shown in Table 2.36. Figure 2.25 is a sample breaker nameplate. Not included in the list are the data for closing and tripping control coils and relay data on where it is connected to. These data are needed in order to retrieve the relay signal to control breaking actions.

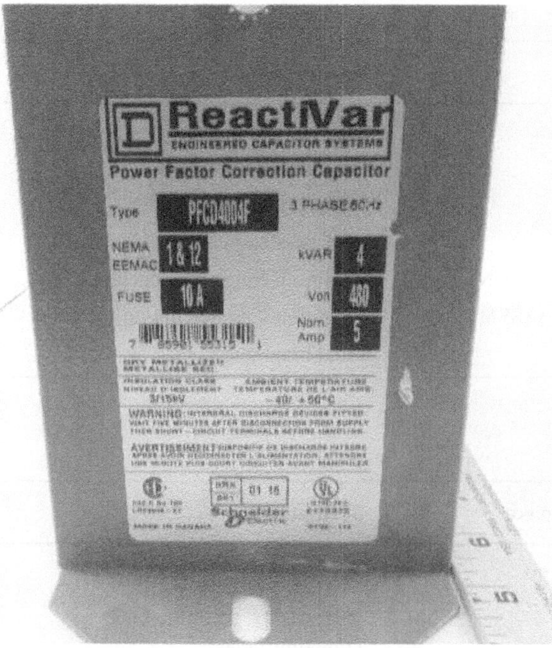

Figure 2.24 Sample capacitor nameplate. Source: Schneider Electric.

Table 2.36 Breaker data.

	Data piece	Description	Note
(a)	Item no.	Equipment name of ID	
(b)	Max. kV	Rated voltage	
(c)	VF_{range}	Rated voltage range factor	
(d)	BIL	Impulse withstand voltage	
(e)	I_{int}	Interruption current capability	
(f)	$I_{c\&l}$	Rated close and latch current capability	
(g)	I_{sc}	Rated short-circuit current	
(h)	Hz	Rated frequency	
(i)	Cycle	Interruption time in cycles	
(j)	T_{delay}	Time delay	

Table 2.37 Fuse data.

	Data piece	Description	Note
(a)	Item no.	Equipment name of ID	
(b)	KV NOM.	Rated nominal voltage	
(c)	KV MAX	Maximum voltage	
(d)	KV BIL	Basic insulation level	
(e)	AMP MAX. (CONT.)	Maximum continuous current	
(f)	AMP INT. (ASYM.)	Asymmetrical interrupting current	
(g)	HZ	Rated frequency	

Table 2.38 Relay data.

	Data piece	Description	Note
(a)	Item no.	Equipment name or ID	
(b)	I_1	Long pickup current	
(c)	t_1	Long time delay	
(e)	I_2	Short pickup current	
(f)	T_2	Short time delay	
(g)	I_3	Instantaneous pickup current	
(h)	T_3	Instantaneous time delay	
(i)	VOLT	Rated voltage	
(j)	CT ratio	Connecting CT ratio	
(k)	CT item No.	CT connected equipment	
(l)	PT ratio	Connecting PT ratio	
(m)	PT item No.	PT connected bus	
(n)	TD	Time dial	

Fuse data. Fuse data are shown in Table 2.37. Figure 2.26 is a sample fuse nameplate.

Relay data. Some over-current relay data are shown in Table 2.38. Figure 2.27 is a sample relay panel.

Busbar data. Some busbar data is shown in Table 2.39. Figure 2.28 is a sample busbar nameplate.

Line data. Some major line data are shown in Table 2.40. Figure 2.29 is a sample line datasheet.

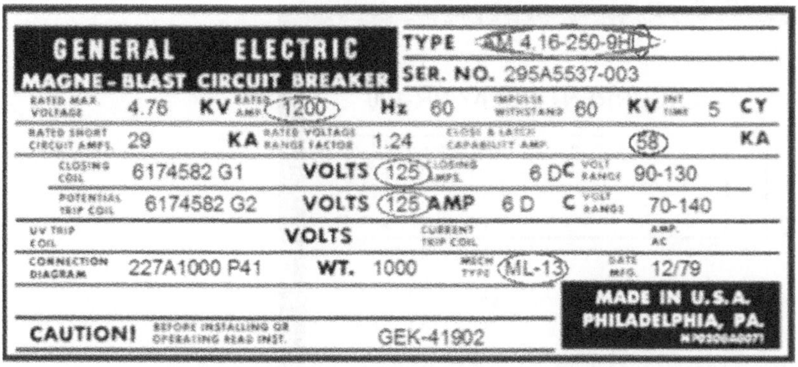

Figure 2.25 Sample breaker nameplate. Source: https://www.google.com/search?q=power+breaker+nameplate&sca_esv=7f561bb85968220c&udm=2&biw=1248&bih=639&ei=BjlGZviBCPidkPIPg5ma-AQ&ved=0ahUKEwj4nLjGwJiFAxX4DkQIHYOMBk8Q4dUDCBA&uact=5&oq=power+breaker+nameplate&gs_lp=Egxnd3Mtd2l6LXNlcnAiF3Bvd2VyIGJyZWFrZXIgbmFtZXBsYXRlSIQ1UABY3yhwAHgAkAEAmAE8oAGEBaoBAjEzuAEDyAEA-AEBmAIHoALxAsICBhAAGAcYHsICCBAAGAUYBxgewgIIEAAYCBgHGB7CAgUQABiABMICBhAAGAUYHsICBxAAGIAEGBiYAwCSBwE3oAegGw&sclient=gws-wiz-serp#vhid=iVYUDwKiqyQ13M&vssid=mosaic/with permission of National Power Equipment, Inc.

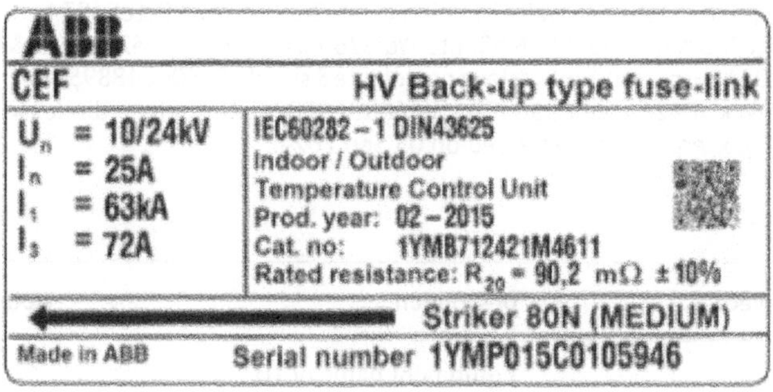

Figure 2.26 Sample fuse nameplate. Source: ABB Inc.

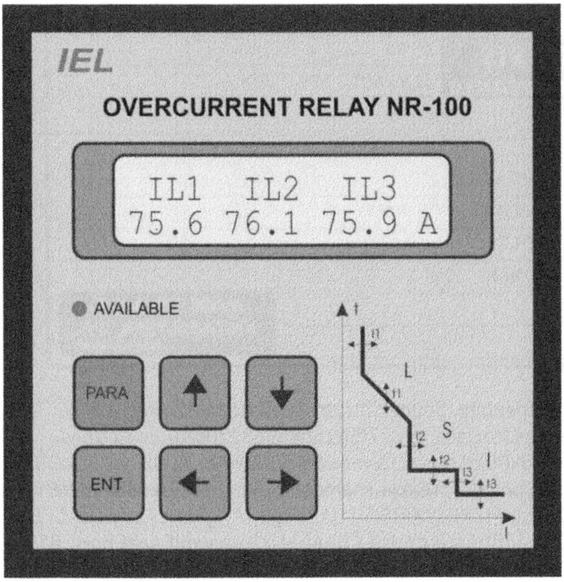

Figure 2.27 Sample relay datasheet/nameplate. Source: Adapted from https://www.bing.com/images/search?view=detailV2&ccid=HoaTTkd9&id=60A7D468B8F6B5D3A18562AADB12041F32A28ED0&thid=OIP.HoaTTkd9kB9MSr1xy5zCiwHaHC&mediaurl=https%3a%2f%2fwww.iel.hr%2fwp-content%2fuploads%2f2018%2f10%2fEN-NR-100-overcurrent-relay.jpg&cdnurl=https%3a%2f%2fth.bing.com%2fth%2fid%2fR.1e86934e477d901f4c4abd71cb9cc28b%3frik%3d0I6iMh8EEtuqYg%26pid%3dImgRaw%26r%3d0&exph=1336&expw=1407&q=overcurrent+relay+nameplate&simid=608006488959841809&FORM=IRPRST&ck=E42AAD6FFD303ED3185608DC8FB04E2C&selectedIndex=6&itb=0&ajaxhist=0&ajaxserp=0/Last accessed on 05 Sep 2024.

Table 2.39 Busbar data.

	Data piece	Description	Note
(a)	Item no.	Equipment name or ID	
(b)	MAX AMP	Maximum allowed current	
(c)	MAX VOLTS	Maximum allowed voltage	

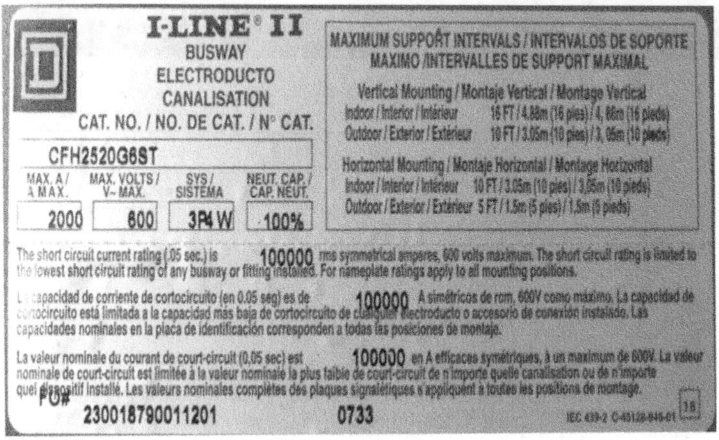

Figure 2.28 Sample busbar nameplate.

Table 2.40 Line data.

	Data piece	Description	Note
(a)	Item no.	Equipment name or ID	
(b)	VOLT	Rated voltage	
(c)	AMP	Rated current per-phase	
(d)	KCMIL	Conductor size	
(e)	# COND	Number of conductors per phase	
(f)	R_{dc}	DC resistance per unit length	
(g)	R_{ac} (25 °C)	AC resistance per unit length	
(h)	R_{ac} (50 °C)	AC resistance per unit length	
(i)	R_{ac} (75 °C)	AC resistance per unit length	
(j)	GMR	Geometric mean radius	
(k)	L	Length	
(l)	CONF	Line layout configuration	
(m)	L	inductance per unit length	
(n)	C	Capacitance per unit length	
(o)	BIL	Basic insulation level	

TABLE A8.1. BARE ALUMINUM CONDUCTORS, STEEL REINFORCED (ACSR) ELECTRICAL PROPERTIES OF MULTILAYER SIZES (*Cont'd*)

				Resistance					Phase-to-Neutral, 60 Hz Reactance at One ft Spacing	
					ac-60 Hz					
Code Word	Size (kcmil)	Stranding Al./St.	Number of Aluminum Layers	dc 20°C (Ohms/ Mile)	25°C (Ohms/ Mile)	50°C (Ohms/ Mile)	75°C (Ohms/ Mile)	GMR (ft)	Inductive Ohms/ Mile X_a	Capacitive Megohm-Miles X_a'
Flicker	477	24/7	2	0.1889	0.194	0.213	0.232	0.0283	0.432	0.0992
Hawk	477	26/7	2	0.1883	0.193	0.212	0.231	0.0290	0.430	0.0988
Hen	477	30/7	2	0.1869	0.191	0.210	0.229	0.0304	0.424	0.0980
Osprey	556.5	18/1	2	0.1629	0.168	0.184	0.200	0.0284	0.432	0.0981
Parakeet	556.5	24/7	2	0.1620	0.166	0.183	0.199	0.0306	0.423	0.0969
Dove	556.5	26/7	2	0.1613	0.166	0.182	0.198	0.0313	0.420	0.0965
Eagle	556.5	30/7	2	0.1602	0.164	0.180	0.196	0.0328	0.415	0.0957
Peacock	605	24/7	2	0.1490	0.153	0.168	0.183	0.0319	0.418	0.0957
Squab	605	26/7	2	0.1485	0.153	0.167	0.182	0.0327	0.415	0.0953
Teal	605	30/19	2	0.1475	0.151	0.166	0.181	0.0342	0.410	0.0944
Kingbird	636	18/1	2	0.1420	0.147	0.162	0.175	0.0301	0.425	0.0951
Rook	636	24/7	2	0.1417	0.146	0.160	0.174	0.0327	0.415	0.0950
Grosbeak	636	26/7	2	0.1411	0.145	0.159	0.173	0.0335	0.412	0.0946
Swift	636	36/1	3	0.1410	0.148	0.162	0.176	0.0300	0.426	0.0964
Egret	636	30/19	2	0.1403	0.144	0.158	0.172	0.0351	0.406	0.0937
Flamingo	666.6	24/7	2	0.1352	0.139	0.153	0.166	0.0335	0.412	0.0943
Crow	715.5	54/7	3	0.1248	0.128	0.141	0.153	0.0372	0.399	0.0920
Starling	715.5	26/7	2	0.1254	0.129	0.142	0.154	0.0355	0.405	0.0928
Redwing	715.5	30/19	2	0.1248	0.128	0.141	0.153	0.0372	0.399	0.0920
Coot	795	36/1	3	0.1146	0.119	0.130	0.142	0.0335	0.412	0.0932
Cuckoo	795	24/7	2	0.1135	0.118	0.128	0.140	0.0361	0.403	0.0917
Drake	795	26/7	2	0.1129	0.117	0.128	0.139	0.0375	0.399	0.0912
Mallard	795	30/19	2	0.1122	0.116	0.127	0.138	0.0392	0.393	0.0904
Tern	795	45/7	3	0.1143	0.119	0.130	0.141	0.0352	0.406	0.0925
Condor	795	54/7	3	0.1135	0.117	0.129	0.140	0.0368	0.401	0.0917
Crane	874.5	54/7	3	0.1030	0.107	0.117	0.127	0.0387	0.395	0.0902
Ruddy	900	45/7	3	0.1008	0.106	0.115	0.125	0.0374	0.399	0.0907
Canary	900	54/7	3	0.1002	0.104	0.114	0.124	0.0392	0.393	0.0898
Corncrake	954	20/7	2	0.0950	0.099	0.109	0.118	0.0378	0.396	0.0898
Rail	954	45/7	3	0.09526	0.0994	0.109	0.118	0.0385	0.395	0.0897
Towhee	954	48/7	3	0.0950	0.099	0.108	0.118	0.0391	0.393	0.0896
Redbird	954	24/7	2	0.0945	0.098	0.108	0.117	0.0396	0.392	0.0890
Cardinal	954	54/7	3	0.09452	0.0983	0.108	0.117	0.0404	0.389	0.0890
Ortolan	1033.5	45/7	3	0.08798	0.0922	0.101	0.110	0.0401	0.390	0.0886
Curlew	1033.5	54/7	3	0.08728	0.0910	0.0996	0.108	0.0420	0.385	0.0878
Bluejay	1113	45/7	3	0.08161	0.0859	0.0939	0.102	0.0416	0.386	0.0874
Finch	1113	54/19	3	0.08138	0.0851	0.0931	0.101	0.0436	0.380	0.0867
Bunting	1192.5	45/7	3	0.07619	0.0805	0.0880	0.0954	0.0431	0.382	0.0864
Grackle	1192.5	54/19	3	0.07600	0.0798	0.0872	0.0947	0.0451	0.376	0.0856

(continued)

Figure 2.29 Sample line datasheet/table. Source: https://www.chegg.com/homework-help/questions-and-answers/appendix-8-data-transmission-line-parameters-table-a81-bare-aluminum-conductors-steel-rein-q92905727.

Table 2.41 Cable data.

	Data piece	Description	Note
(a)	Item no.	Equipment name or ID	
(b)	Size	Nominal cross section area of conductor	mm^2 or in^2
(c)	Diameter	Outer diameter	mm or in
(d)	Weight	Weight per unit length	kg/km or lb/mile
(e)	VOLT	Rated voltage	Volts
(f)	R_{dc}	DC resistance per unit length at 20 °C	Ω/km
(g)	L	Inductance per unit length	mH/km
(h)	C	Capacitance per unit length	μF/km
(i)	I_{grd}	Underground installation current rating	Amp
(j)	I_{air}	Open air current rating	Amp
(k)	I_{sc}	5s short-circuit current rating	kA
(l)	BIL	Basic insulation level	kV

Table 2.42 Starting device data.

	Data piece	Description	Note
(a)	Item no.	Equipment name or ID	
(b)	Type	Starting device type	
(c)	Rating	Starting device ratings	
(d)	Connection	Starting device connection	
(e)	Sensor	Starting device sensor connection	
(f)	Control	Starting device control function and parameters	

Cable data. Some major cable data are shown in Table 2.41. Figure 2.30 is a sample cable datasheet.

Starting device data. Motor starting devices vary and their data is provided in various formats. Some major data are shown in Table 2.42. Figure 2.31 is a sample starting device configuration.

Harmonic source data. Harmonic sources are typically represented as harmonic spectrum. Figure 2.32 are harmonic data in spectrum format and Table 2.43 lists what should be entered into computer models.

U0/U (Um): 19/33 (36) kV **Cu/XLPE/CTS/PVC/SWA/PVC**

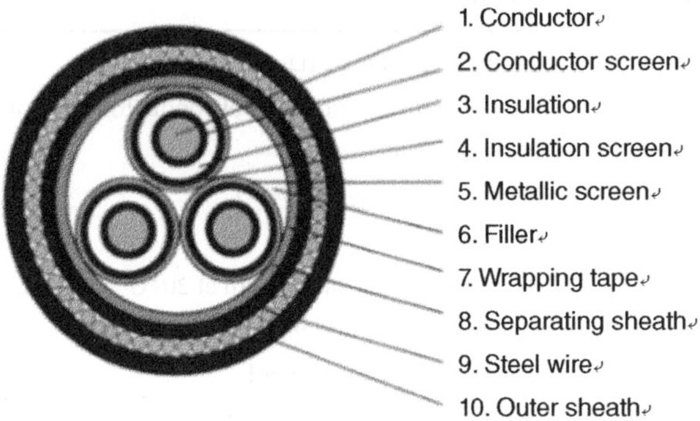

Nom. cross sect areas of conductor mm^2	Structure parameters		Electrical parameters			Current rating of cable		Conductor current at short circuit (5S) kA
	Outer diameter (approx.) mm	Weight (approx.) kg/km 76.72	Max conductor D.C. resistance at 20°C ≤ Ω/km	Conductor inductance mH/km	Operating capacitance μF/km	Under ground (20°C)	In Air (30°C)	
50	76.72	9173	0.387	0.3963	0.1363	181	205	3.2
70	80.78	10296	0.268	0.3761	0.1508	220	253	4.48
95	84.63	11558	0.193	0.3605	0.1643	263	307	6.08
120	88.04	12742	0.153	0.349	0.176	298	352	7.67
150	91.69	14083	0.124	0.3376	0.1893	332	397	9.59
185	96.14	15966	0.0991	0.3278	0.2026	374	453	11.83
240	101.91	18453	0.0754	0.3154	0.2207	431	529	15.35
300	106.61	20846	0.0601	0.3067	0.2387	482	599	19.19
400	113.66	24288	0.047	0.296	0.2648	541	683	25.58

Figure 2.30 Sample cable datasheet. Source: https://pdf.directindustry.com/pdf/sunway-company/xlpe-insulated-power-cable-rated-voltage-19-33kv/235306-949405.html/with permission of VirtualExpo/Last accessed on 05 Sep 2024.

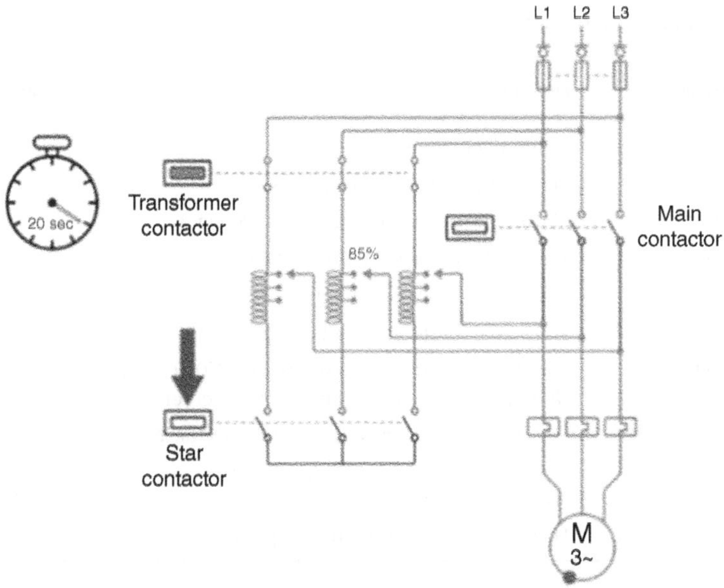

Figure 2.31 Sample starting device configuration. Source: https://www.realpars.com/blog/motor-starter/with permission of RealPars B.V./Last accessed on 05 Sep, 2024.

Harmonic filter data. Harmonic filter data include filter type (structure) and parameters. Table 2.44 lists the general data requirements for harmonic filters. For detailed filter type and structure, please refer to section 3.1.10 Harmonic filter. Figure 2.33 shows a general harmonic filter data template.

System configuration. I&C power system configuration defines the exact topology and connections of each equipment, normally represented by single-line diagram. Single-line diagram is a single-phase presentation of a balanced three-phase system, generally adapted by the industry. If there are unbalanced phases including sources, circuits, and loads, it needs special clarifications and additional data specifications. The single-line diagrams nowadays are generated by computer software by providing equipment list and connection information of each equipment. The digital single-line diagram can include a database to keep model and parameter information of the entire system to be used for computer aided studies and analysis. Table 2.45 specifies configuration data needed to create an I&C power system single-line diagram. Figure 2.34 demonstrates a sample single-line diagram of I&C power system showing buses, branches and connected equipment with their names and IDs.

Table 2.43 Harmonic source data.

	Data piece	Description	Note
(a)	Item no.	Equipment name or ID	
(b)	Bus	Harmonic source connected bus	
(c)	Type	Harmonic source type (current or voltage)	
(d)	VOLT	Harmonic source operating voltage	
(e)	F	Harmonic source operating frequency	
(f)	kVA	Harmonic source apparent power rating	
(g)	Spectrum	Harmonic spectrum data series: $\{u_{mag}(h), u_{ang}(h)\}$ ($h = 1, \ldots, h_{max}$ - maximum harmonic order)	
(h)	GRD CONN	Harmonic source phase connection (Delta, Y)	
(i)	GRD	Harmonic source grounding method (grounded, open)	
(j)	X_{grd}	Harmonic source grounding impedance	

Table 2.44 Harmonic filter data.

	Data piece	Description	Note
(a)	Item no.	Equipment name or ID	
(b)	Bus	Filter connected bus	
(c)	VOLT	Filter operating voltage	
(d)	F	Filter operating frequency	
(e)	VAR	Filter reactive power at fundamental frequency	
(f)	Type	Filter type	
(g)	CONF	Filter configuration diagram	
(h)	PARA	Filter parameter list: $\{L, C, R\}$	
(i)	Hz	Filter tuning frequency (frequencies): $\{f_1, \ldots, f_n\}$	
(j)	I_{sc}	Short-circuit current	
(k)	BIL	Basic insulation level	
(l)	$V_{w/s}$	Short time withstand voltage	
(m)	GRD CONN	Harmonic filter phase connection (Delta, Y)	
(n)	GRD	Harmonic filter grounding method (grounded, open)	
(o)	X_{grd}	Harmonic filter grounding impedance	

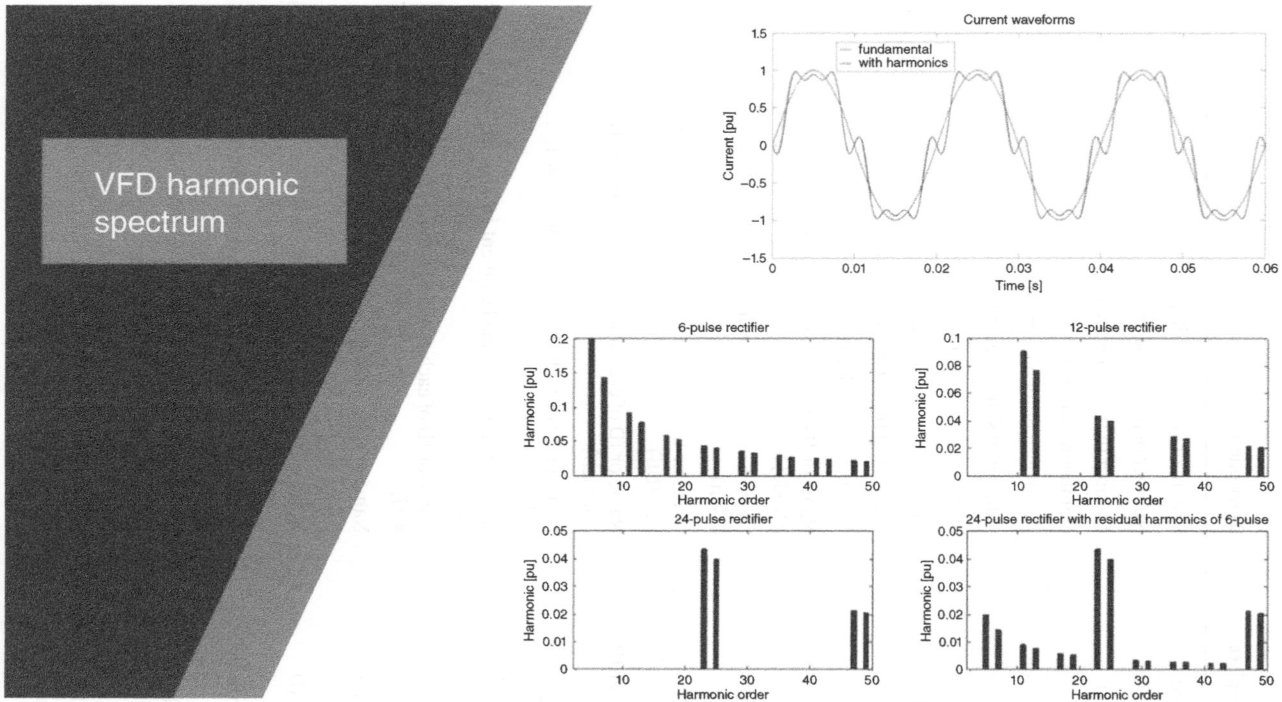

Figure 2.32 Sample harmonic source data. Source: https://mb-drive-services.com/network_harmonics_vfd_harmonic_spectrum//MB Drive Services/Last accessed on 06 Sep 2024.

Table 2.45 Configuration data.

	Data piece	Description	Note
(a)	Name	System or single-line diagram name or ID	
(b)	#Node	Number of nodes and buses	
(c)	#Gen	Number of source elements	
(d)	#Uti	Number of utility connections	
(e)	#Branch	Number of branch elements	
(f)	#Shunt	Number of shunt elements	
(g)	#Load	Number of load elements	
(h)	#Pd	Number of protective devices	
(i)	#Tap	Number of transformers with tap	
(j)	#Cap	Number of switchable capacitors	
(k)	NodeID	Name or ID of each bus and node n ($n = 1, \ldots,$ #Node)	
(l)	GenID	Name or ID of each source element g (g = 1, …, #Gen)	
(m)	UtiID	Name or ID of each utility connection u ($u = 1, \ldots,$ #Uti)	
(n)	BranchID	Name or ID of each branch element b ($b = 1, \ldots,$ #Branch)	
(o)	ShuntID	Name or ID of each shunt element s ($s = 1, \ldots,$ #Shunt)	
(p)	LoadID	Name or ID of each load element l ($l = 1, \ldots,$ #Load)	
(q)	PdID	Name or ID of each protective device p ($p = 1, \ldots,$ #Pd)	
(r)	TapID	Name or ID of each transformer tap t ($t = 1, \ldots,$ #Tap)	
(s)	CapID	Name or ID of each capacitor switch tap c ($c = 1, \ldots,$ #Cap)	
(t)	GenNode	Connecting node or bus name or ID of each generation element g (g = 1, …, #Gen)	
(u)	UtiNode	Connecting node or bus name or ID for each Utility connection u ($u = 1, \ldots,$ #Uti)	
(v)	BranchNode1, BranchNode2	From and to connecting node or bus names or IDs of each branch element b ($b = 1, \ldots,$ #Branch)	
(w)	ShuntNode	Connecting node or bus name or ID of each shunt element s ($s = 1, \ldots,$ #Shunt)	

Table 2.45 (Continued)

	Data piece	Description	Note
(x)	LoadNode	Connecting node or bus name or ID of each load element l ($l = 1, \ldots, \#\text{Load}$)	
(y)	PdNode or PdElement	Connecting bus or element name or ID of each protective device n ($n = 1, \ldots, \#\text{Pd}$)	
(z)	ElementStatus	Operating condition of each generation, branch, shunt and load element e ($e = 1, \ldots, \#\text{Gen} + \#\text{Branch} + \#\text{Shunt} + \#\text{Load}$)	
(aa)	PdStatus	Status of each protective device p ($p = 1, \ldots, \#\text{Pd}$)	
(bb)	ElementConn	Phase connection of each generation, shunt and load element ec ($ec = 1, \ldots, \#\text{Gen} + \#\text{Shunt} + \#\text{Load}$)	
(cc)	ElementGrd	Grounding method of each generation, shunt and load element eg ($eg = 1, \ldots, \#\text{Gen} + \#\text{Shunt} + \#\text{Load}$)	

RATING	RANGE OF AVAILABLE RATINGS
Bank Configuration:	Fixed, Switched, Single Step, Multiple Step
Filter Types:	Notch (Band-Pass), High-Pass, C-High-Pass, Double-Tuned, Multi-Tuned
Operating Voltage (line-to-line):	2.4kV - 38kV
Operating Frequency:	50 Hertz | 60 Hertz
Reactive power output:	0.5MVAR - 100 MVAR (500kvar - 100,000 kvar)
Tune frequency (Hz)	85 Hz 2100 Hz (500kvar - 100,000 kvar)
High-Pass (damping) resistor rating:	1 ohm to 1000 ohms 10kW/Phase - 200kW/Phase
Short circuit (asymmetrical momentary):	16kA - 61kA
Impulse withstand voltage (Basic Insulation Level):	60kV - 200 KV
Short-time withstand voltage (1 minute 50/60 Hertz):	19kV - 100kV

Figure 2.33 Sample harmonic filter datasheet. Source: Adapted from https://nepsi.com/products/metal-enclosed-harmonic-filter-banks/#ratings/Last accessed on 05 Sep,2024

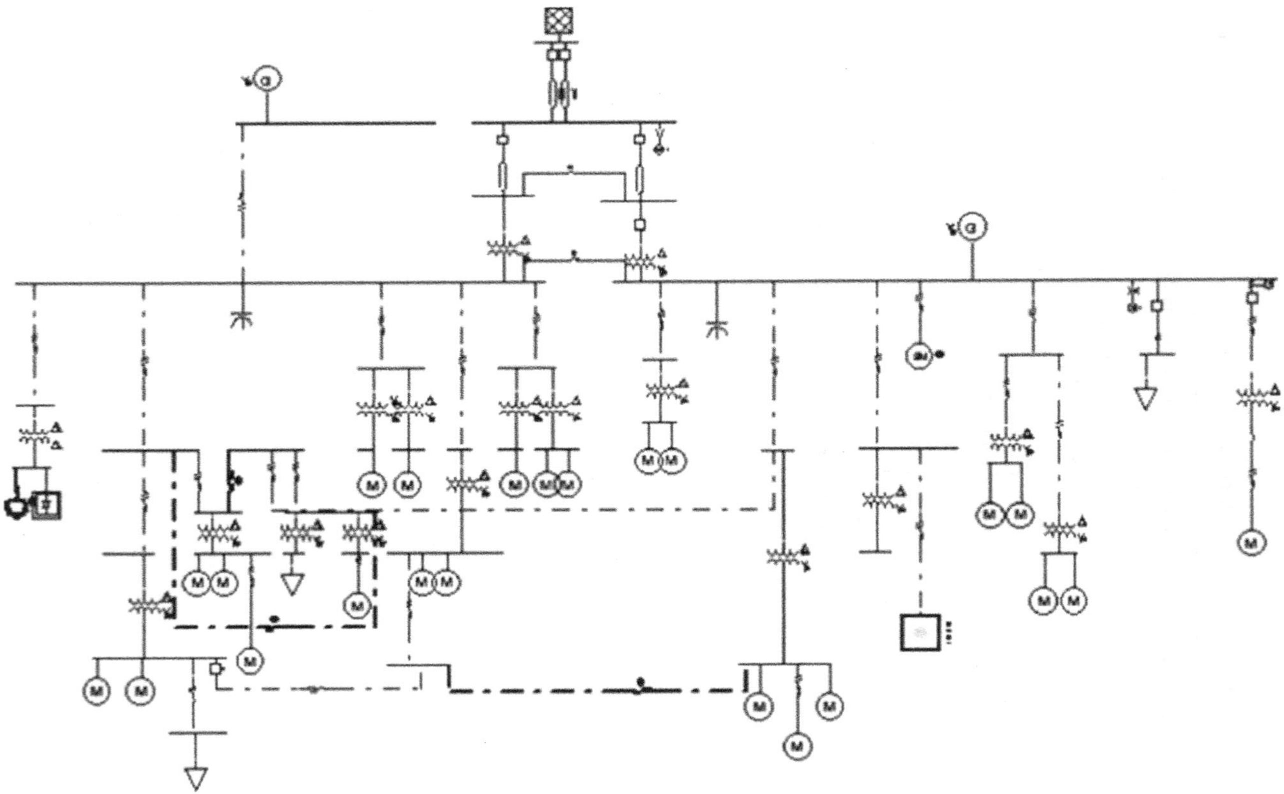

Figure 2.34 Sample I&C power system configuration.

Table 2.46 Study data.

	Data piece	Description	Note
(a)	Study	Study case/scenario name or ID	
(b)	Std	Industry standard(s) and related parameters	
(c)	Temp	Temperature	
(d)	Unit	Unit system	
(e)	LdFile	Initial loading profile	
(f)	GenFile	Initial generation profile	
(g)	VFile	Initial operating voltage profile	
(h)	UVFile	Initial utility voltage profile	
(i)	LdStatus	Initial load status profile	
(j)	GenStatus	Initial generation status profile	
(k)	UStatus	Initial utility status profile	
(l)	UVvar	Utility voltage variation range	
(m)	Vvar	Voltage variation tolerance	
(n)	Fvar	Frequency variation tolerance	
(o)	DTlim	Device coordination minimum time	
(p)	RAlim	Rotor angle swing limit	
(q)	LFSch	Load change schedule	
(r)	SCSch	Short-circuit location schedule	
(s)	TCCSch	TCC zone schedule	
(t)	StartMSch	Starting motor schedule	
(u)	HarSch	Harmonic source schedule	
(v)	EventSch	Transient stability study event schedule	
(w)	#LF	Number of load flow configurations	
(x)	#Fault	Number of locations for fault duty calculation	
(y)	#TCC	Number of TCC curves	
(z)	#Device	Number of devices for each TCC t ($t = 1, \ldots, \#TCC$)	
(aa)	#StartingMotor	Number of starting motors	
(bb)	#Motor	Number of motor plots for motor starting	
(cc)	#Bus	Number of bus plots for motor staring	
(dd)	#Event	Number of events in the study case	
(ee)	#RotorA	Number of rotor angle plots	
(ff)	#Freq	Number of frequency plots	
(gg)	#Volt	Number of voltage plots	
(hh)	#MotorS	Number of motor speed plots	

Table 2.47 Load flow analysis output data.

	Data piece	Description	Note
(a)	NodeV	Voltage magnitude with phase angle for each node and bus n ($n = 1, \ldots$, #Node)	
(b)	BranchP, BranchQ, BranchI	Branch active and reactive power and current flow for each branch element b ($b = 1, \ldots$, #Branch)	
(c)	BranchVdrop	Branch voltage drop for each branch element b ($b = 1, \ldots$, #Branch)	
(d)	BranchPloss, BranchQloss	Branch active and reactive power losses for each branch element b ($b = 1, \ldots$, #Branch)	
(e)	GenP, GenQ, GenI	Active, reactive power output and current for each generation element g ($g = 1, \ldots$, #Gen)	
(f)	GenV	Terminal voltage for each generation element g ($g = 1, \ldots$, #Gen)	
(g)	UtiP, UtiQ, UtiI	Utility importing/exporting active and reactive power and current for each utility connection u ($u = 1, \ldots$, #Uti)	
(h)	LoadP, LoadQ	Load active and reactive power assumption for each load element l ($l = 1, \ldots$, #Load)	
(i)	LoadV, LoadI	Voltage and current for each load element l ($l = 1, \ldots$, #Load)	
(j)	TapPos	Transformer tap position for each transformer tap t ($t = 1, \ldots$, #Tap)	
(k)	CapPos	Capacitor switching position for each switching capacitor c ($c = 1, \ldots$, #Cap)	
(l)	TotalP, TotalQ	Total system active and reactive loading	
(m)	TotalGenP, TotalGenQ	Total system active and reactive power generation	
(n)	TotalPloss, TotalQloss	Total system active and reactive power losses	
(o)	#VFlag	Number of voltage violations for node and bus	
(p)	#GFlag	Number of reactive power violations for generation element	
(q)	#BFlag	Number of current violations for branch element	
(r)	#SFlag	Number of voltage violations of shunt element	
(s)	#LFlag	Number of current violations for load elements	

Table 2.47 (Continued)

	Data piece	Description	Note
(t)	VFlag	Bus and node name or ID for violation *bf* (*bf* = 1, … , #VFlag)	
(u)	GFlag	Generation element name or ID for violation *gf* (*gf* = 1, … , #GFlag)	
(v)	BFlag	Branch element name or ID for violation *bf* (*bf* = 1, … , #BFlag)	
(w)	SFlag	Shunt element name or ID for violation *sf* (*sf* = 1, … , #SFlag)	
(x)	LFlag	Load element name or ID for violation *lf* (*lf* = 1, … , #LFlag)	

Table 2.48 Short-circuit analysis output data.

	Data piece	Description	Note
(a)	#CbFlag	Number of circuit breaker capacity violations	
(b)	#FuseFlag	Number of fuse capacity violations	
(c)	I_{mom}	Momentary fault duties at specified locations f ($f = 1, \ldots,$ #Fault)	
(d)	I_{int}	Interrupting fault duties at specified locations f ($f = 1, \ldots,$ #Fault)	
(e)	I_{ss}	Steady state fault duties at specified locations f ($f = 1, \ldots,$ #Fault)	
(f)	V_{node}	Voltage drops at nodes and buses n ($n = 1, \ldots,$ #Node)	
(g)	CbFlag	Circuit breaker name or ID with violation *cb* (*cb* = 1, … , #CbFlag)	
(h)	FuseFlag	Fuse name or ID with violation *fu* (*fu* = 1, … , #FuseFlag)	

Table 2.49 Motor starting analysis output data.

	Data piece	Description	Note
(a)	#VFlag	Number of voltage violations	
(b)	#AFlag	Number of motor acceleration failures	
(c)	#VNode	Number of node and bus voltage monitoring	
(d)	I_{start}	Motor starting current curve *sm* (*sm* = 1, …, #StartingMotor)	
(e)	T_{start}	Motor starting torque curve *sm* (*sm* = 1, …, #StartingMotor)	
(f)	V_{start}	Motor starting voltage curve *sm* (*sm* = 1, …, #StartingMotor)	
(g)	S_{start}	Motor starting speed/slip curve *sm* (*sm* = 1, …, #StartingMotor)	
(h)	VFlag	Bus name or ID for voltage violation *vf* (*vf* = 1, …, #VFlag)	
(i)	AFlag	Motor name or ID for acceleration failure *af* (*af* = 1, …, #SFlag)	
(j)	V_{Node}	Node or bus voltage for each monitoring location *vn* (*vn* = 1, …, # VNode)	

2.4.2 Study Data

Study data provides the information related to a specific study case or scenario. Typical study data include industry standard to be complied for the current study case, system initial operating condition, study related parameters, output specifications, and some software tool specific variables. Table 2.46 lists some commonly used study data.

2.4.3 Output Data

Output data provides calculation results for the specified analysis type under the given study scenarios. These data are organized in various formats that are easy to read and understand, offering rich information on the system performance, highlighting findings and violations according to the defined requirements, or serving as benchmarked records. Tables 2.47–2.52 list typical key output data arranged based on study and analysis types.

Table 2.50 Harmonic analysis output data.

	Data piece	Description	Note
(a)	#PCCVFlag	Number of PCC voltage harmonic distortion violations	
(b)	#PCCIFlat	Number of PCC current harmonic distortion violations	
(c)	#VFlag	Number of node or bus harmonic voltage distortion violations	
(d)	#IFlag	Number of branch current harmonic distortion violations	
(e)	#PRFlag	Number of parallel resonances	
(f)	#SRFlag	Number of serial resonances	
(g)	#Filter	Number of filters	
(h)	V_{thd}	Node and bus voltage total harmonic distortion indices i (i = 1, …, #Node)	
(i)	V_{ihd}(h)	Node and bus voltage individual harmonic distortion indices i (i = 1, …, #Node, h = 2, …, 50)	
(j)	I_{thd}	Branch current total harmonic distortion indices b (b = 1, …, #Branch)	
(k)	I_{ihd}(h)	Branch current individual harmonic distortion indices b (b = 1, …, #Branch, h = 2, …, 50)	
(l)	PCCVFlag	PCC name or ID with voltage harmonic distortion violation pv (pv = 1, …, #PCCVFlag)	
(m)	PCCIFlag	PCC name or ID with current harmonic distortion violation pi (pi = 1, …, #PCCIFlag)	
(n)	VFlag	Bus and node name or ID with voltage harmonic distortion violation vf (vf = 1, …, #VFlag)	
(o)	IFlag	Branch name or ID with current harmonic distortion violation if (if = 1, …, #IFlag)	
(p)	PRFlag	Bus and node name or ID with parallel resonance pr (pr = 1, …, #PRFlag)	
(q)	PRPoint	Frequency and magnitude for each parallel resonance point pr (pr = 1, …, # PRFlag)	
(r)	SRFlag	Branch name or ID with serial resonance sr (sr = 1, …, #SRFlag)	
(s)	SRPoint	Frequency and magnitude for each serial resonance point sr (sr = 1, …, # SRFlag)	
(t)	FilterF	Tuning frequency dataset for each filter f (f = 1, …, #FilterF)	
(u)	FilterPara	Parameter dataset for each filter p (p = 1, …, #FilterF)	

Table 2.51 Protection and device coordination output data.

	Data piece	Description	Note
(a)	#CoordFlag	Number of device coordination violations	
(b)	#ThermFlag	Number of equipment thermal damage violations	
(c)	#MtrXfmFlag	Number of starting motor and energizing transformer inrush violations	
(d)	MinTReq	Minimum time coordination requirement	
(e)	Setting	Setting dataset for each device d in TCC t ($d = 1, \ldots,$ #Device; $t = 1, \ldots,$ #TCC)	
(f)	MinT	Minimum time coordination for each TCC t ($t = 1, \ldots,$ #TCC)	
(g)	CoordFlag	Coordination violation device name or ID dv ($dv = 1, \ldots,$ #CoordFlag)	
(h)	ThermFlag	Thermal damage violation equipment name or ID tv ($tv = 1, \ldots,$ #ThermFlag)	
(i)	MtrXfmFlag	Starting or inrush violation motor or transformer name or ID mv ($mv = 1, \ldots,$ #MtrXfmFlag)	

Table 2.52 Transient stability analysis output data.

	Data piece	Description	Note
(a)	#RotorAFlag	Number of rotor angle stability violations	
(b)	#FreqFlag	Number of frequency variation violations	
(c)	#VoltFlag	Number of voltage variation violations	
(d)	#MotorSFlag	Number of motor speed drop violations	
(e)	#RelayTrip	Number of relay trips	
(f)	RotorALim	Rotor angle limit	
(g)	FreqLim	Frequency limit	
(h)	VoltLim	Voltage limit	
(i)	MotorSLim	Motor speed limit	
(j)	RotorA	Rotor angle plots ra ($ra = 1, \ldots,$ #RotorA)	
(k)	Freq	Frequency plots f ($f = 1, \ldots,$ #Freq)	
(l)	Volt	Voltage plots v ($v = 1, \ldots,$ #Volt)	
(m)	MotorS	Motor speed plots ms ($ms = 1, \ldots,$ #MotorS)	
(n)	RotorAFlag	Rotor angle violation element name or ID rav ($rav = 1, \ldots,$ #RotorAFlag)	
(o)	FreqFlag	Frequency violation element name or ID fv ($fv = 1, \ldots,$ #FreqFlag)	
(p)	VoltFlag	Voltage violation element name or ID vv ($vv = 1, \ldots,$ #VoltFlag)	
(q)	MSFlag	Motor starting violation element name or ID msv ($msv = 1, \ldots,$ #MotorSFlag)	
(r)	RelayTrip	Relay trip element name or ID rt ($tr = 1, \ldots,$ #RelayTrip)	

References

1 IEEE Std 399, IEEE Recommended Practice for Industrial and Commercial Power Systems Analysis (IEEE Brown Book), IEEE, 1997.

2 IEEE Std 141, Recommended Practice for Electric Power Distribution for Industrial Plants, IEEE, 1999.

3 Prabhakara, F.S. and Smith, R.L., R.L. Smith, JR (1995). *Industrial and Commercial Power Systems Handbook*. McGraw-Hill.

4 J.R. Dunki-Jacobs, C.E. Davis, "The Value of Conceptual Power System Design Studies," Annual Technical Conference on Pulp and Paper Industry Proceedings, 18–22 June 1990. (https://ieeexplore.ieee.org/document/109875)

5 V.J. Mishra, M.D. Khardenvis, 2012 IEEE Students' Conference on Electrical, Electronics and Computer Science, 1–2 March 2012, Bhopal, India, 2012. (https://ieeexplore.ieee.org/document/6184751)

6 IEC 60034–1, Rotating Electrical Machines - Part 1: Rating and Performance, IEC, 2022.

7 IEEE Std 519, IEEE Standard for Harmonic Control in Electric Power Systems, IEEE, 2022.

8 IEEE Std 18, IEEE Standard for Shunt Capacitors, IEEE, 2012.

9 Dai, J.J. and Shokooh, F. (Jan.-Feb. 2022). "Industrial and Commercial Power System Harmonic Studies: Introduction to IEEE Standard 3002.8-2018". *IEEE Industry Applications Magazine* 28 (1): 51–64.

10 ANSI/NETA, Standard for Maintenance Testing Specifications for Electrical Power Equipment and Systems, NETA, 2023.

11 Dai, J.J. (Jan.-Feb. 2024). "A Complete Electrical System Study for a Refinery With Major Expansion". *IEEE Transactions on Industry Applications* 60 (1): 1599–1610.

12 NFPA 70E, Standard for Electrical Safety in the Workplace, National Fire Protection Association (NFA), 2024.

13 IEEE Std 1584, IEEE Guide for performing Arc-Flash Hazard Calculations, IEEE, 2018.

3

Computer Modeling of I&C Power Systems

To analyze industrial and commercial (I&C) power systems, modeling of the entire system with each component in different levels of details and complexities must be conducted first. Computer modeling is a powerful tool for understanding, designing, and analyzing complex responses and interactions of these components with I&C power systems. It provides rich information based on industrial standards and matured numerical techniques to help engineers to assess the system performance, derive improvements, and confirm enhancements so the entire electrical infrastructure and operation procedures can be more reliable, economical, and stable. Through modeling, the entire system, including equipment is transformed into a digital twin and represented by appropriate mathematical equations that will be solved by various algorithms and methods by computers. Computer-simulated and solved solutions in digital format are then arranged into engineering reports for design or operation engineers to review and evaluate. In this chapter, the most fundamental modeling requirements for I&C power systems, equipment modeling, and network modeling will be discussed and illustrated. Other analyses associated with modeling requirements for the control system and operation mode will also be covered briefly.

3.1 Equipment and Component Modeling

Basically, every electrical equipment and component involved in power system operation and performance needs to be properly modeled in mathematical form. These models will be used by computer software tools to represent, simulate, and analyze the equipment and component behavior, performance, and interactions with the electrical system. The models are essential for designing, optimizing, and troubleshooting power systems. The fundamental modeling form is the relationship between basic electrical quantities measured at the terminal of the

Industrial and Commercial Power System Analysis Fundamentals and Practice, First Edition. J. J. Dai.

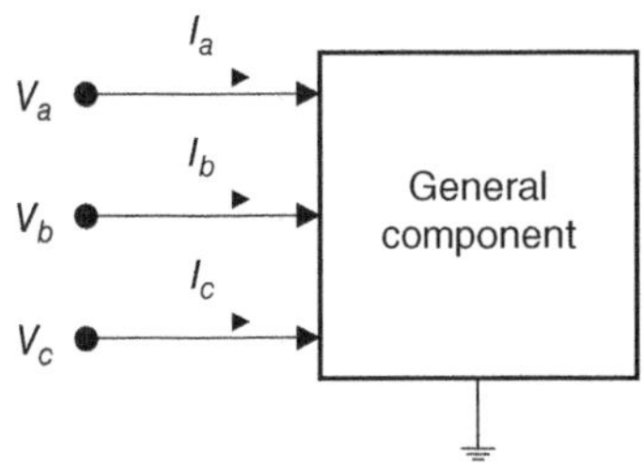

Figure 3.1 General electrical component model.

equipment, namely, voltage, current, and power. The model's internal parameters are sources, if any, impedance, and admittance, sometimes including time constant, time delay, and others. In I&C power systems, most equipment and component models are linear or have been linearized to fit into analyzing algorithms. The general I&C power system electrical component model is illustrated in Figure 3.1 with three-phase voltage-current inputs.

Following common and practical I&C power system equipment and components are considered, and their corresponding models for load flow, short-circuit, motor starting, harmonic, and transient stability analyses are discussed. Due to the symmetry of the three-phase voltages, all models are represented by the single-phase equivalent circuit, unless in some special cases with explanations.

3.1.1 Utility

Relative to I&C power systems, utility usually is a rigid power source with large short-circuit capability, nearly fixed voltage and frequency, and relatively constant equivalent impedance. Thus, the utility model for various analyses is straightforward.

Load flow analysis model. For load flow analysis, utility, in general, is modeled as a reference machine with fixed voltage magnitude and phase angle, as shown in Table 3.1.

The reference machine is a modeling and computation requirement by digital simulation algorithms. The reference machine is connected to the slack bus.

Table 3.1 Utility modeling for load flow analysis.

	Equipment type	Active power	Reactive power	Voltage magnitude	Voltage phase angle	Notes
(a)	Utility	Variable	Variable	Fixed	Fixed	Voltage regulator and reference machine (slack bus)

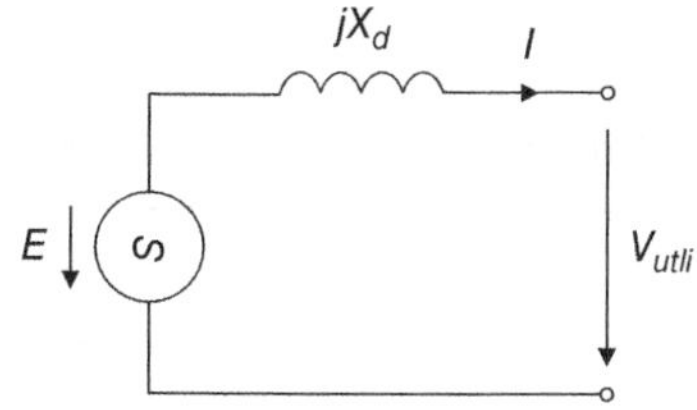

Figure 3.2 Utility for short-circuit analysis model.

Table 3.2 Utility reactance for short-circuit analysis.

Type	Medium-voltage and high-voltage	Low voltage
First-cycle		
Remote utility	$1.0 \times X_s$	$1.0 \times X_s$
Interrupting time (1.5 to 4 cycles)		
Remote utility	$1.0 \times X_s$	∞

Note 1: X_s is system equivalent impedance.
Note 2: 1.5 to 4 cycles interrupting times do not apply to low-voltage circuit breakers.
Source: Adapted from IEEE Std 3002.3-2018 [1].

Short-circuit analysis model. In short-circuit analysis, utility, in general, is modeled as a constant voltage source E behind an equivalent system impedance X_d (Thevenin equivalent) as shown in Figure 3.2 and Table 3.2 which is reproduced from [1] with permission from IEEE. The magnitude of the equivalent internal voltage source is set to the prefault voltage at the utility connecting point.

Motor starting analysis model. When performing motor starting analysis, utility, in general, can be modeled in a similar way as in short-circuit analysis given in Figure 3.2 and Table 3.2 with the equivalent internal voltage as the utility operating voltage and the equivalent impedance as the equivalent system impedance X_s.

Harmonic analysis model. If a utility is not considered a harmonic voltage or current source, then it can be modeled in harmonic analysis using the same model structure as in short-circuit analysis. However, if a utility has harmonic contents that will affect the local I&C power system quality, then the utility should be modeled as a harmonic source with harmonic spectrum specified. Although, in most cases, utility harmonics are predominately modeled as voltage sources, they could be modeled as current sources as well depending on the nature of harmonic effects. Harmonic source modeling will be covered later.

When the utility is involved in harmonic analysis, it is also needed to consider frequency dependency, which means the impedance in the Thevenin equivalent of Figure 3.2 no longer holds as a constant. The impedance should be modeled, including the frequency effects, as shown in Table 3.3.

Table 3.3 Sample harmonic dependent impedance spectrum.

Harmonic order	Resistance	Reactance
2	$R(2)$	$X(2)$
3	$R(3)$	$X(3)$
4	…	…
5	…	…
6	…	…
7	…	…
…	…	…

Transient stability analysis model. It is assumed that compared to the average I&C power system, utilities can be modeled as a constant voltage source behind the system equivalent impedance similar to short-circuit analysis as in Figure 3.2 and Table 3.2. The utility will be assumed to have a constant frequency, too. If this assumption is not true, then utilities should be modeled as an equivalent synchronous machine, as discussed next. Please note that in such a case, utilities rarely are modeled with equivalent excitation and automatic voltage regulator (AVR) and/or prime mover/speed governor systems.

3.1.2 Synchronous Machine

Synchronous machines include generators, condensers, and motors. They are key equipment in I&C power systems. On-site installed synchronous generators provide essential energy to the facility for processes and electrical loads to enhance system reliability, reduce operation costs, and support special needs. Synchronous condensers supply reactive power to improve voltage drop and power factor. Synchronous motors offer special features to drive various loads. Computer modeling of synchronous machines is a crucial aspect of power system analysis because they are particularly important to system frequency, voltage, and stability performance. Modeling of synchronous machines in I&C power systems involves creating mathematical representations of the machines as well as their control systems to allow simulating their behavior under different operating conditions. Based on the study type, synchronous machines are modeled differently.

Load flow analysis model. Synchronous machines are connected to the network through terminal buses as shown in Figure 3.3. Different synchronous machines are modeled and treated differently in load flow analysis. Table 3.4 summarizes the model specifications. Synchronous generators are essential power sources to provide active and reactive power, as well as to regulate voltage in load flow analysis.

Based on the operation requirements, a synchronous generator can be designed and configured in three different modes as shown in Table 3.4 rows (a), (b), and(c), where modes (a) and (b) corresponding to the generator speed governor in the droop setting, and (c) to the isochronous setting. Synchronous condensers only provide reactive power, and they can regulate voltage, too, since it has an excitation system. Synchronous motors are constant power loads that withdraw both active and reactive power from the system.

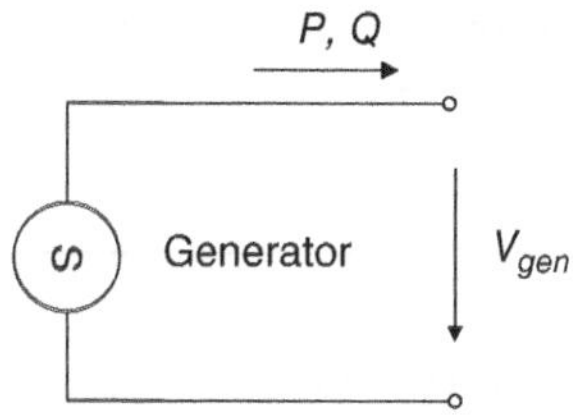

Figure 3.3 Synchronous machine model.

Short-circuit analysis model. The equivalent circuit as in Figure 3.2 also applies to synchronous machines, i.e. a simplified calculation approach with variable reactance plus an equivalent internal voltage. Table 3.5 shows the variable reactance values for different types of synchronous machines depending on their relative electrical distance to the fault location. There are several IEEE standards relevant to short-circuit analysis and synchronous and induction machine modeling. Readers are advised to find out details about these standards that are applicable to their particular system conditions [2–6].

Motor starting analysis model. When it is reasonable to assume during motor starting, generator voltage regulators have had time to increase field excitation sufficiently to maintain the desired generator terminal voltage or are electrically remote to the starting motors, then Figure 3.2 and Table 3.5 can be used to represent operating generators. Otherwise, a more realistic model of Figure 3.4 should

Table 3.4 Synchronous generator operation modes for load flow analysis.

	Machine type	Active power	Reactive power	Voltage magnitude	Voltage phase angle	Notes
(a)	Generator	Fixed output	Variable	Regulator	Variable	Active power provider and voltage regulator
(b)	Generator	Fixed output	Fixed output	Variable	Variable	Active and reactive power provider
(c)	Generator	Variable	Variable	Fixed	Fixed	Voltage regulator and reference machine (slack bus)
(d)	Condenser	None	Fixed output	Regulator	Variable	Reactive power provider and voltage regulator
(e)	Motor	Fixed input	Fixed input	Variable	Variable	Active and reactive power load

Table 3.5 Synchronous machine reactance for short-circuit analysis.

Type	Medium voltage and high voltage	Low voltage
First cycle		
Local generator	$1.0 \times X_s$	$1.0 \times X_s$
Synchronous motor	$1.0 \times X''_{dv}$	$1.0 \times X''_{dv}$
Interrupting-time calculations (1.5 to 4 cycles)		
Local generator	$1.0 \times X''_{dv}$	∞
Synchronous motor	$1.5 \times X''_{dv}$	∞

Note 1: X''_{dv} direct-axis saturated subtransient reactance at rated voltage.
Note 2: 1.5 to 4 cycles interrupting times do not apply to low-voltage circuit breakers.
Source: Adapted from IEEE Std 3002.3-2018 [1].

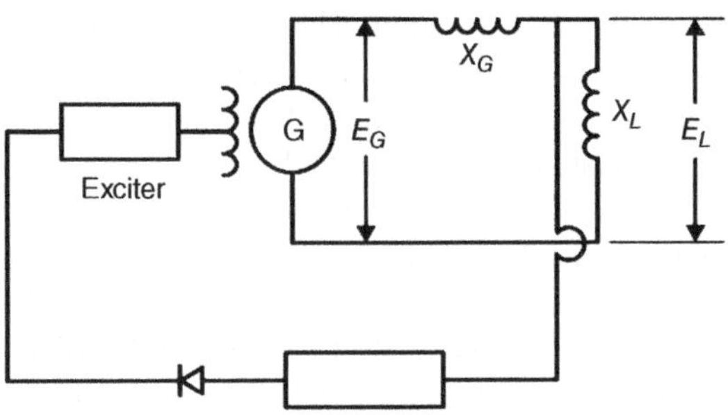

Figure 3.4 Synchronous generator model for motor starting analysis. Source: IEEE Std 3002.7-2018 [7]/with permission of IEEE.

be adopted in which the voltage applied to a load by a locally connected generator is a function of time.

Variables in Figure 3.4 are found as: [7]

$E_L = E_G \frac{X_L}{X_L + X_G}$ is the generator voltage.

E_G is generator internal voltage due to excitation and varies with time as $T''_{do} \rightarrow T'_{do} \rightarrow T_{do}$.

X_G is generator internal reactance and varies with time as $T''_d \rightarrow T'_d \rightarrow T_d$.

X_L is load reactance.

T''_d, T'_d, T_d are generator subtransient, transient, and steady-state time constants on the d-axis, respectively.

$T''_{do}, T'_{do}, T_{do}$ are generator subtransient, transient, and steady-state open-circuit time constants on the d-axis, respectively.

Harmonic analysis model. For harmonic analysis purposes, synchronous machines are modeled as rotating machine equivalent, which is a harmonic resistance R_h and harmonic reactance X_h equivalent in series, as shown in Figure 3.5.

Resistance and reactance in Figure 3.5 are given by: [8]

$$R_h = R\sqrt{h} \tag{3.1a}$$

$$X_h = hX_2 \text{ or } X_h = \frac{(X_d'' + X_q'')}{2} \tag{3.1b}$$

where:

h is harmonic order,

R is machine resistance derived from the machine power loss at the fundamental frequency,

X_2, X_d'', X_q'' are machine negative sequence reactance, d-axis subtransient, and q-axis subtransient reactance, receptively.

Transient Stability Analysis Model. Full dynamic models need to be used to represent synchronous machines to simulate the machine's behavior during transient events, such as faults, sudden load changes, loss of utility connections, and other system impacts. Figure 3.6 illustrates a synchronous machine circuit diagram that includes original stator *abc* three-phase windings and a rotor field winding. Using Park transformation, the standard method to describe synchronous machine equations, the *abc* three-phase circuits are transformed to *dq* circuits, so time-varying variables are avoided in the model [9]. The model can include additional damper windings on the field winding circuit, i.e. the *d*-axis and on the *q*-axis (only one damping winding on each of the *d* and *q* axes is shown).

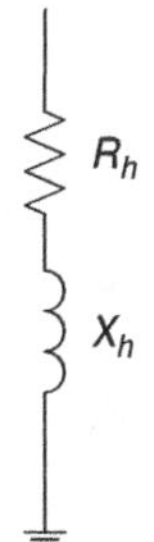

Figure 3.5 Synchronous machine model for harmonic analysis. Source: IEEE Std 3002.8-2018 [8]/with permission of IEEE.

Various synchronous machine equivalent circuit models can be derived based on the number of damping windings to be included. Table 3.6 lists selected circuit models per IEEE Standard 1110-2019 [10].

These equivalent circuit models involve differential equations to describe the machine's response over the time domain. Taking Model 2.1 as an example, the corresponding equations can be developed as,

$$\left[\frac{dI}{dt}\right] = [L_2]^{-1}\{-[R] - \omega[L_1]\}[I] + [L_2]^{-1}[V] \tag{3.2}$$

where vectors $[V]$ and $[I]$ representing voltages and currents defined as:

$$[V] = [v_d\, v_q\, e_{fd}\, v_{1d}\, v_{1q}]^T$$

$$[I] = [i_d\, i_q\, i_{fd}\, i_{1d}\, i_{1q}]^T$$

and $[R]$, $[L_1]$, $[L_2]$ are relative resistance and inductance matrices.

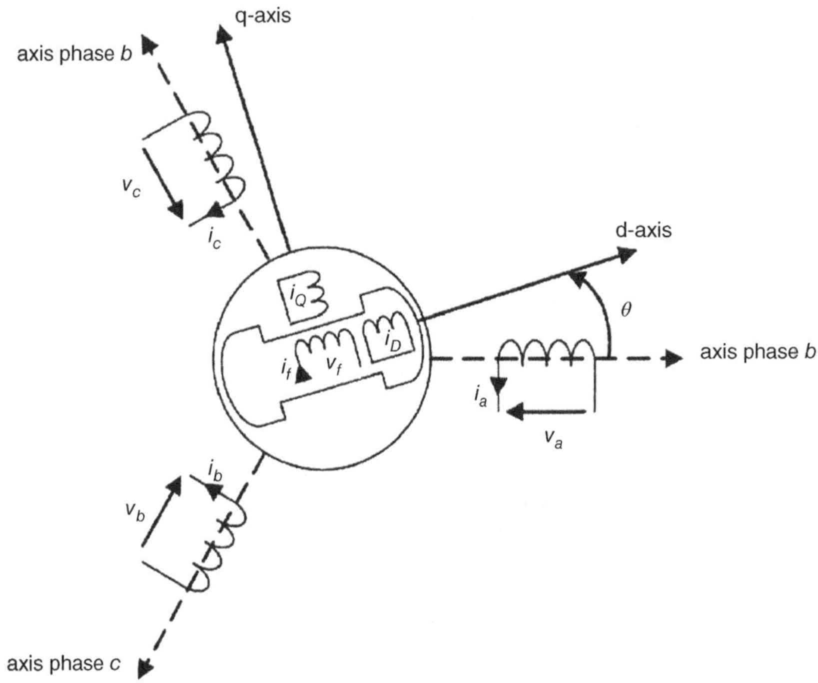

Figure 3.6 Synchronous machine transient analysis model. Source: Dai [9]/ with permission of IGI Global.

A complete solution of the synchronous machine during transients should include a mechanical model to express the motion equation for the rotor shaft. This equation is derived from Newton's second law, and it is referred to as the swing equation in power system transient analysis [9]:

$$J\frac{d\omega}{dt} = T_m - T_e - D\omega \tag{3.3}$$

where

ω is the machine angular speed.
J is the angular moment of inertia of the machine rotor shaft.
D is the damping constant.
T_m is the mechanical torque applied on the machine rotor shaft.
T_e is electromechanical torque applied on the machine rotor shaft.

The mechanical torque T_m is provided by the machine prime mover, which is modeled separately. The electromechanical torque T_e is generated interactions between machine winding currents and winding flux linkages.

Table 3.6 Selected synchronous machine dynamic models.

Constant rotor flux linkages		Thevenin equivalent 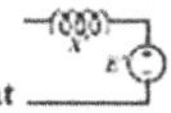		
Q-axis / **D-axis**	**No equivalent damper circuit**	**One equivalent damper circuit**	**Two equivalent damper circuit**	**Three equivalent damper circuit**
Field circuit only	Model 1.0 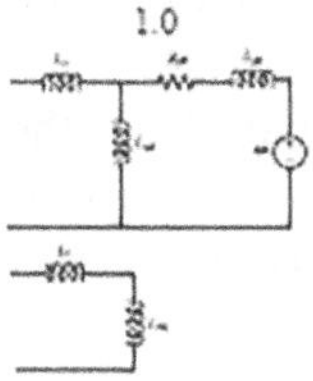	Model 1.1 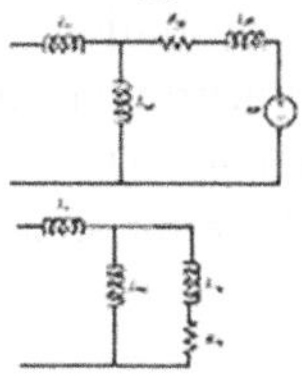	Not considered	Not considered
Field circuit + one equivalent damper circuit	Not considered	Model 2.1 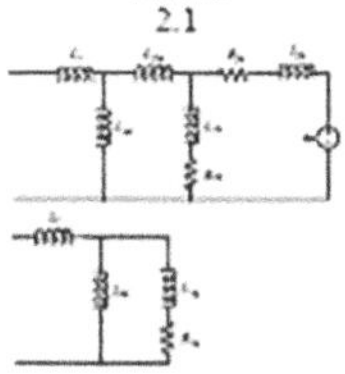	Model 2.2 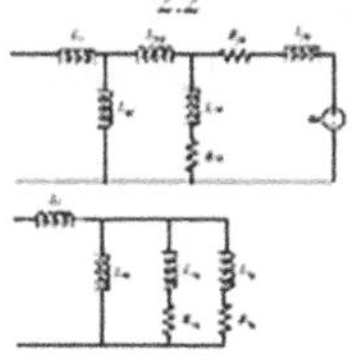	Model 2.3 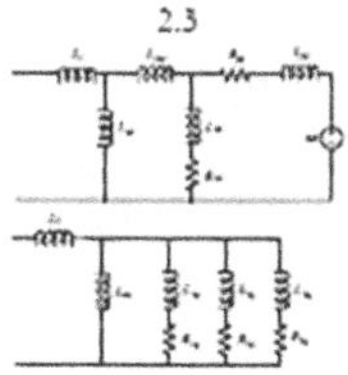
Field circuit + two equivalent damper circuits	Not considered	Not considered	Not considered	Model 3.3 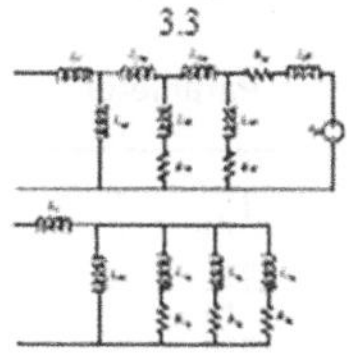

Source: IEEE1110-2019 [10]/with permission of IEEE.

If neglect the damping and substitute the torque by power, the swing equation can be written as,

$$M\frac{d\omega}{dt} = P_m - P_e \tag{3.4}$$

where

M is the machine inertia constant,
P_m is the mechanical power applied on the machine rotor shaft, and
P_e is the machine's electrical power output.

Machine rotor angle δ which is defined as the angular displacement between the rotor's magnetic field axis and the stator's magnetic field axis can be obtained from,

$$\delta = \frac{d\omega}{dt} - \delta_0 \tag{3.5}$$

The schematic of power and torque balances in the swing equation is illustrated in Figure 3.7.

One set of sample synchronous machine dynamic model data typically provided by manufacturers and vendors is listed in Table 3.7 which is reproduced from [10] with permission from IEEE for demonstration purposes.

Synchronous condenser and synchronous motor have the same machine equations as the synchronous generator. The only differences between them are the mechanical engagements at the rotor shaft, i.e. the generator has a prime

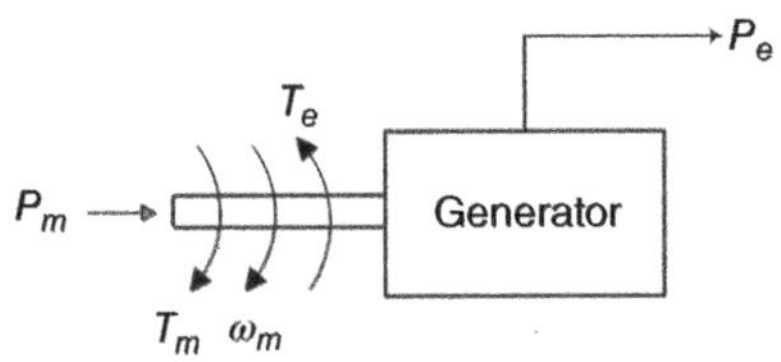

Figure 3.7 Power and torque balance schematic.

Table 3.7 Typical unsaturated parameters of synchronous machine.

Parameter	Machine size: 310 MVA/ 13.8 kV, 60 Hz/128.6 rpm.
T'_{do}	6.6 sec.
T'_d	2.2 sec.
T''_{do}	0.05 sec
T''_d	0.046 sec.
T_a	0.26 sec.
T''_{qo}	0.10 sec.
X'_d	0.33 pu
X''_d	0.25 pu
X''_q	0.33 pu
X_d	1.14 pu
X_q	0.63 pu

Source: Adapted from IEEE1110-2019 [10].

mover input to provide power, the condenser has no mechanical input, and the motor has a mechanical load. Prime movers and motor loads will be addressed in the following sections of this chapter, as well as excitation systems that are a part of the integrated system for synchronous machines.

3.1.3 Induction Machine

Induction machines play a very important role in I&C power systems, and their modeling is probably next to that of synchronous machines in terms of complexity. Induction machines include both induction generators and induction motors. Their models for different system studies are described below.

Load flow analysis model. Induction machines are modeled in load flow analysis as either a power source for generator operation, a constant power load for the running motor, or a constant impedance load for the starting motor, as shown in Table 3.8.

Short-circuit analysis model. Induction machines can be represented by the Thieving equivalent shown in Figure 3.8. Different from synchronous machines, during faults, induction machine internal voltage will decay faster with time due to lack of excitation winding. Alternatively, this circuit model can be assumed to be in such a mode that the equivalent reactance will increase by multipliers based on the machine ratings and operating conditions. Values of the suggested variable equivalent reactance with multipliers for induction machine in short-circuit analysis are suggested in Table 3.9 which is reproduced from [1] with permission from IEEE.

Motor starting analysis model. A starting induction motor can be modeled using an equivalent circuit model as shown in Figure 3.9. In the model, R_1 and R_2 represent motor stator winding and rotor winding resistance, and X_1, X_2, and X_m

Table 3.8 Induction machine operation modes for load flow analysis.

	Machine type	Active power	Reactive power	Impedance	Notes
(a)	Generator	Fixed output	Determined by terminal voltage	n/a	Active power provider and reactive power load
(b)	Running motor	Fixed input	Determined by terminal voltage	n/a	Active and reactive power load
(c)	Starting motor	n/a	n/a	Fixed impedance	Impedance load

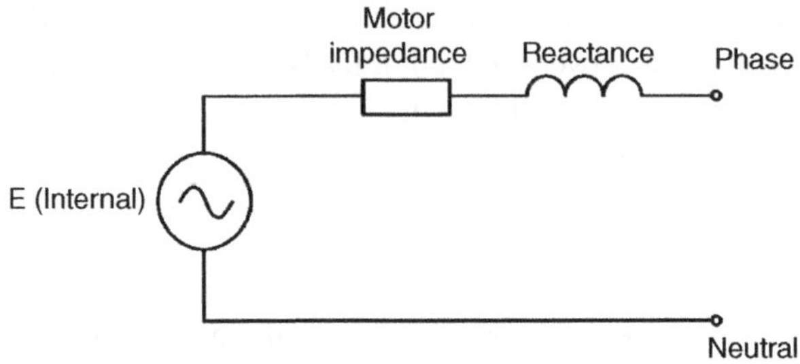

Figure 3.8 Induction machine short-circuit analysis model. Source: IEEE Std 3002.3-2018 [1]/with permission of IEEE.

Table 3.9 Induction machine reactance multiplying factor.

Machine specification	**Medium voltage and high voltage**	**Low voltage**
First-cycle calculations		
Large size (above 1000 HP, or above 250 HP and 2-pole)	$1.0X''$	$1.66X''$
Medium (50 to 249 HP, or 250 to 1000 HP more than 2-pole)	$1.2X''$	$1.66X''$
Small (Below 50 HP)	∞	$1.66X''$
Interrupting-time calculations (three to five cycles)		
Large (Above 1000 HP, or above 250 HP and 2-pole)	$1.5X''$	n/a
Medium (50 to 249 HP, or 250 to 1000 HP more than 2-pole)	$3.0X''$	n/a

Source: Adapted from IEEE Std 3002.3-2018 [1].

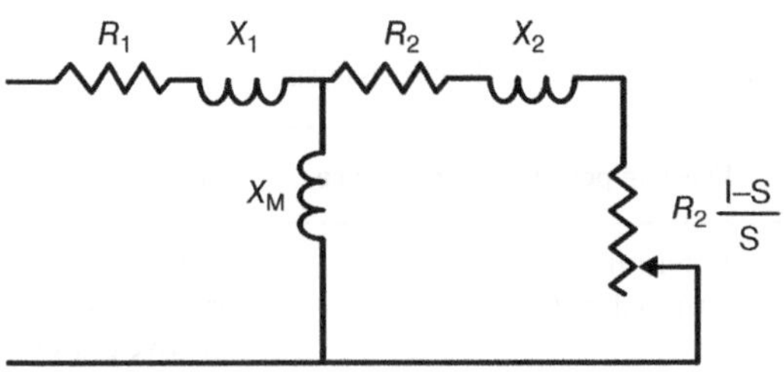

Figure 3.9 Induction machine for motor starting analysis simplified model. Source: IEEE Std 3002.7-2018 [7]/ with permission of IEEE.

represent motor stator winding, rotor winding, and air gap reactance, respectively. S is the motor speed slip. When the motor terminal voltage is given, motor stator and rotor currents can be calculated, as well as motor torque, both are functions of the motor slip.

Assuming a three-phase motor, the electrical torque on the motor shaft can be computed as: [11]

$$T_e = 3\frac{p_f}{2}\frac{R_2}{s\omega_s}I_2^2 \tag{3.6}$$

where

p_f is the number of poles.
$\omega_s = 2\pi f$ is the synchronous speed.
I_2 is the rotor current.

Figure 3.9 is a simplified model without considering the double-cage rotor structure, rotor with deep bars, and other sophisticated build induction machines. There are some high-degree equivalent circuit models available, too. If requiring improved accuracy, machine characteristic curves either from validated manufacturer data or field testing should be used. These curves include machine speed versus current, speed versus power factor, and speed versus torque, as shown in Figure 3.10.

Harmonic analysis model. Induction machines for harmonic studies and analysis will use the general rotating machine harmonic model shown in Figure 3.5 and Equation (3.1).

Transient stability analysis model. For transient stability studies and analysis in I&C power systems, the induction machine model in Figure 3.9 can provide fairly

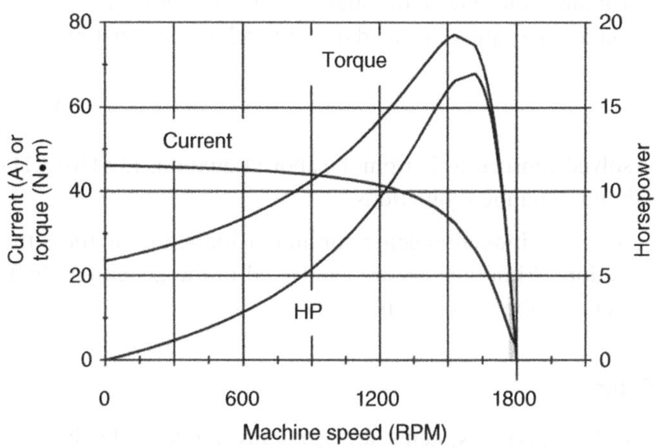

Figure 3.10 Induction machine for motor starting analysis characteristic curve model. Source: Starting Methods of Three Phase Induction Motor / with permission of Electrical Academia.

seasonable results, so it is wildly adopted. If more detailed fast transient phenomena are preferred, the following complex variable equations or d–q models in [11] can be used.

Assume the induction machine is modeled by a Thevenin equivalent circuit of a complex transient voltage $\overline{E}'$ behind the stator resistance R_1 and a transient reactance X' where X' is calculated based on Figure 3.9:

$$X' = X_1 + \frac{X_2 X_m}{X_2 + X_m} \tag{3.7}$$

The equation for machine terminal voltage $\overline{V}$, machine stator current $\overline{I}_1$ and differential equation for $\overline{E}'$ in complex variable form can be written as,

$$\overline{V} = \overline{I}_1 + (R_1 + jX')\overline{I}_1 \tag{3.8}$$

$$\frac{d\overline{E}'}{dt} = -j\omega_s s\overline{E}' - \frac{1}{T_0'}(\overline{E}' - j(X_0 - X')\overline{I}_1) \tag{3.9}$$

where

$X_0 = X_1 + X_m$ is rotor open-circuit reactance.
$T_0' = \frac{X_2 + X_m}{\omega_s R_2}$ is rotor open-circuit time constant.

By including the differential equation for slip,

$$\frac{ds}{dt} = \frac{1}{2H}(T_m - T_e) \tag{3.10}$$

where

H is the inertia constant.
T_m is the machine mechanical torque on the shaft from the motor load.
T_e is the machine's electrical torque calculated from the following equation:

$$T_e = \frac{Re(\overline{E}'^{*}\overline{I}_1)}{\omega_s} \tag{3.11}$$

$\overline{I}_1, \overline{E}', s, T_e$ are to be solved consecutively from the above Equations (3.8) to (3.11) in the time domain for dynamic simulations.

The difference between an induction generator and an induction motor is the machine shaft connection. Prime movers are connected to the generator shaft, while loads are connected to the motor shaft.

3.1.4 Line and Cable

Lines and cables in I&C power systems are generally modeled by lumped parameters in pi circuit as shown in Figure 3.11. This model is valid for load flow, short-circuit, motor starting, and transient stability studies and analyses.

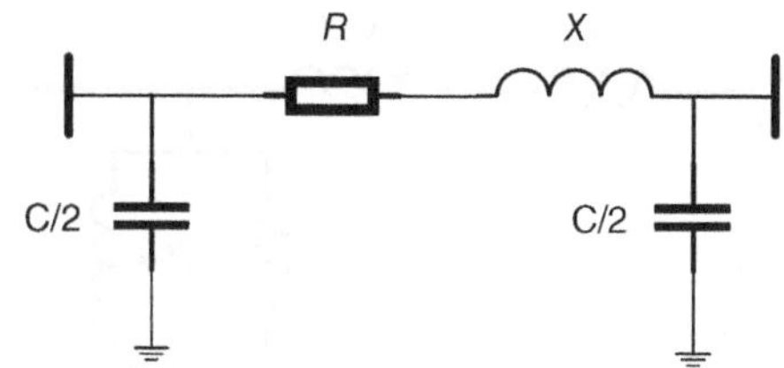

Figure 3.11 Line and cable model for system analysis.

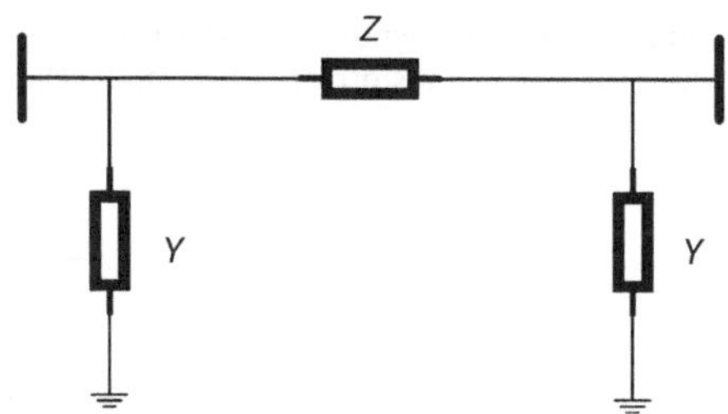

Figure 3.12 Long-line model for system analysis. Source: IEEE Std 3002.8-2018 [8]/with permission of IEEE.

Harmonic analysis model. For harmonic analysis, since frequency can be high, thus, the waveform propagation effect cannot be neglected. A long-line model should be used as shown in Figure 3.12.

Line frequency dependent series impedance Z and shunt admittance Y in Figure 3.12 for the long-line model are defined below [8]:

$$Z = Z_c \sinh(\gamma_e l) \tag{3.12a}$$

$$\frac{Y}{2} = \frac{1}{Z_c} tanh\left(\frac{\gamma_e}{2} l\right) \tag{3.12b}$$

where

$$Z_c = \sqrt{\frac{z}{y}}$$

$$\gamma_e = \sqrt{zy}$$

$z = r + jx_L$ (Ω/m) = line series impedance at fundamental frequency
$y = g + jb_C$ (S/km) = line shunt admittance at fundamental frequency
l = length of the line (m)

3.1.5 Transformer

Load flow, short-circuit, motor starting, transient stability analyses model. A complete circuit model for a two-winding transformer, with P, S, and M indicating

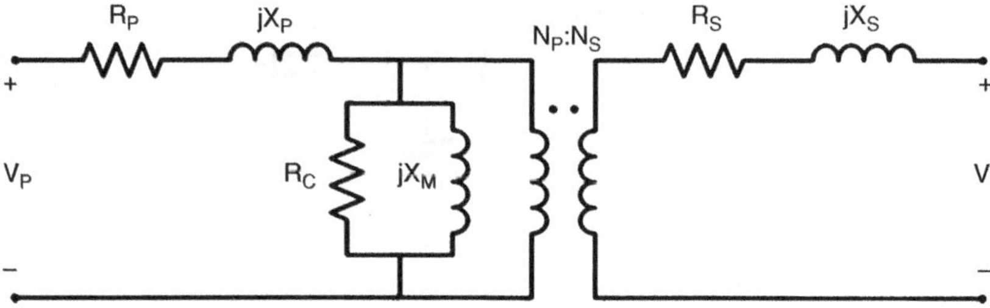

Figure 3.13 Two-winding transformer model. Source: Transformer circuit model Diagram" with "Transformer circuit model Diagram / with permission of Microsoft.

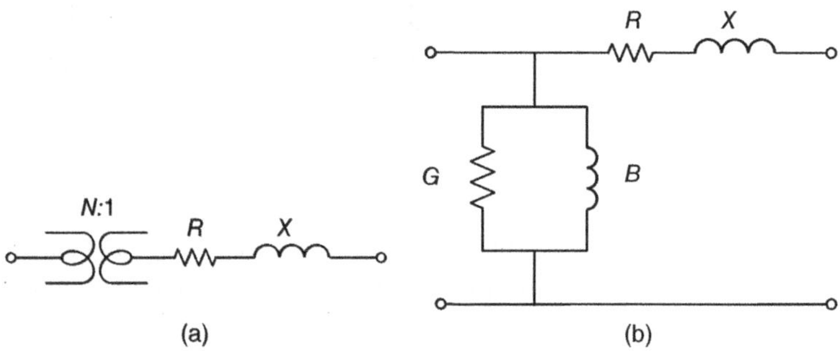

Figure 3.14 Two-winding transformer model for system analyses except for harmonic analysis. (a) Simplified model. (b) Model with magnetizing branch.

primary, secondary, and magnetizing circuits, is shown in Figure 3.13. Simplified versions are found in Figure 3.14 which are valid for load flow, short-circuit, motor starting, and transient stability analyses.

Harmonic analysis model. For harmonic analysis, the model impedances are frequency dependent.

In Figure 3.14,

$R + jX = Z_t$ = Transformer winding impedance.

N = Transformer turn ratio.

$\frac{1}{G+jB} = Z_m$ = Transformer magnetizing impedance.

The three-winding transformer is modeled in three circuits, as shown in Figure 3.15.

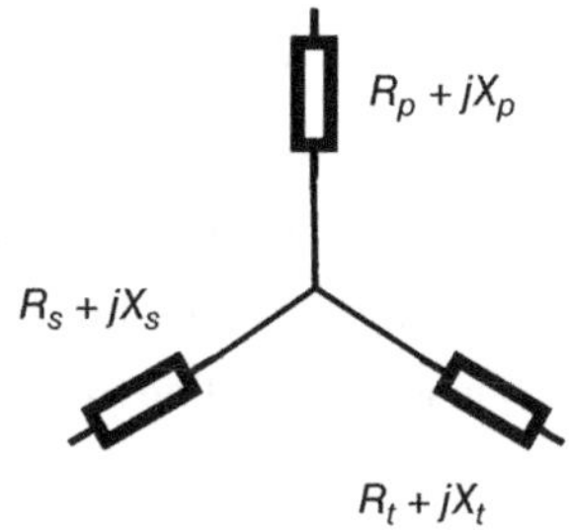

Figure 3.15 Three-winding transformer model for system analyses except for harmonic analysis.

Terminal-to-terminal impedances in Figure 3.15 are derived from the measured impedances below:

$Z_{ps} = R_p + R_s + j(X_p + X_s)$ = Leakage impedance between primary and secondary windings measured with tertiary winding open circuit.

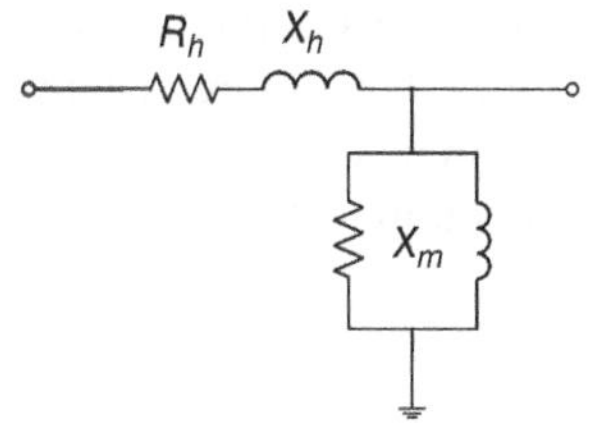

Figure 3.16 Two-winding transformer model for harmonic analysis. Source: IEEE Std 3002.8-2018 [8]/with permission of IEEE.

$Z_{pt} = R_p + R_t + j(X_p + X_t)$ = Leakage impedance between primary and tertiary windings measured with the secondary winding open circuit.

$Z_{st} = R_s + R_t + j(X_s + X_t)$ = Leakage impedance between secondary and tertiary windings measured with the primary winding open circuit.

Harmonic analysis. Due to their frequency contents in voltage and current, transformers should be modeled with frequency-dependent parameters for harmonic analysis, as shown in Figure 3.16 and Equation (3.13). It is noted that the magnetizing shunt branch can be ignored if the transformer is not a significant harmonic source.

The series resistance R_h and reactance X_h at harmonic order h are computed from:

$$R_h = R_T\sqrt{h} \tag{3.13a}$$

$$X_h = hX_T \tag{3.13b}$$

where

R_T is derived from transformer power loss at the fundamental frequency.

X_T is transformer short-circuit reactance.

3.1.6 Generic Load

Generic loads represent a wide range of I&C power system equipment and devices that solely consume or convert powers from the system, including but not limited to lighting, heating and cooling devices, electrical boilers, arc furnaces, computer and IT equipment, and other electrification processes. They are also called passive loads. Generic loads are in shunt connection to network buses, as shown in Figure 3.17. Based on the electrical characteristics, the generic loads can be modeled in different ways.

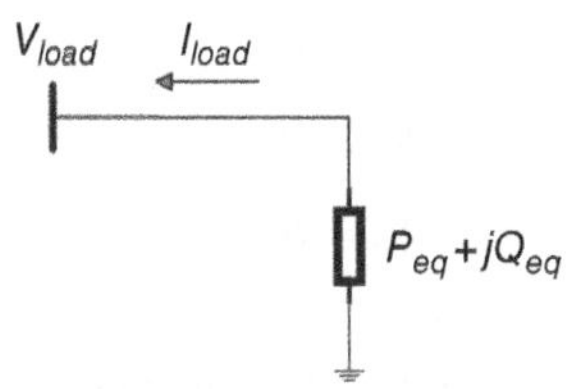

Figure 3.17 Generic load model.

Load flow, short-circuit, motor starting, transient stability analyses model. Some loads present themselves as constant impedance, meaning their power consumptions are proportional to the square of the magnitude of the applied voltage. While some other loads exhibit constant power behavior, meaning their power consumptions are voltage- and current-independent, holding at constant values. For a more general case, load P and Q can be a polynomial function of terminal voltage and load current. These three types of loads can be modeled by Equations (3.14), (3.15), or (3.16):

$$P_{eq} = K_p V^2 \tag{3.14a}$$

$$Q_{eq} = K_q V^2 \tag{3.14b}$$

or

$$P_{eq} = P_{const} \tag{3.15a}$$

$$Q_{eq} = Q_{const} \tag{3.15b}$$

or

$$P_{eq} = P_0 + P_{v1}V + P_{i1}I + P_{v2}V^2 + P_{i2}I^2 + \cdots \tag{3.16a}$$

$$Q_{eq} = Q_0 + Q_{v1}V + Q_{i1}I + Q_{v2}V^2 + Q_{i2}I^2 + \cdots \tag{3.16b}$$

Short-circuit analysis model. Short-circuit analysis focuses on the equipment and devices that contribute to fault currents. Since passive loads are not fault current sources, therefore they are not considered in short-circuit analysis.

Harmonic analysis. Generic loads are sometimes modeled as frequency-dependent impedance in harmonic analysis, as shown in Figure 3.18 and Equations (3.17) and (3.18). The model includes an equivalent resistance R_h in series with an equivalent reactance X_h which are given by,

Figure 3.18 Load model for harmonic analysis. Source: IEEE Std 3002.8-2018 [8]/with permission of IEEE.

$$R = \frac{V^2}{P} \tag{3.17a}$$

$$X = \frac{V^2}{Q} \tag{3.17b}$$

$$R_h = R\sqrt{h} \tag{3.18a}$$

$$X_h = hX \tag{3.18b}$$

where

R and X are equivalent load resistance and reactance.

P and Q are active and reactive power of the load.

3.1.7 Motor Load

Motor loads are the loads that are connected to motor shafts.

Load flow, short-circuit, and harmonic analyses model. For these studies and analyses, sufficient information has already been included in the machine model, thus there is no need to model the motor loads separately.

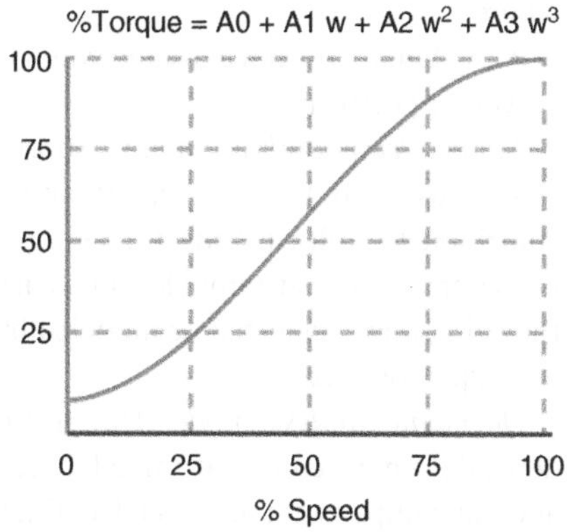

Figure 3.19 A sample load torque-speed curve. Source: IEEE Std 3002.7-2018 [7]/with permission of IEEE.

Motor starting and transient stability analyses model. Motor load describes load mechanical torque versus motor speed and is represented in a torque–speed curve form shown in Figure 3.19 which are required for the starting motor in motor starting analysis, and for both large size running and starting motors in transient stability analysis. This model is applicable to both synchronous motors and induction motors.

The motor load curve is typically described by a polynomial function in Equation (3.19):

$$\text{Torque} = A_0 + A_1 \times \omega + A_2\omega^2 + A_3\omega^3 + \cdots \tag{3.19}$$

where ω is motor speed and As are constant coefficients.

3.1.8 Capacitor and Static Var Compensator (SVC)

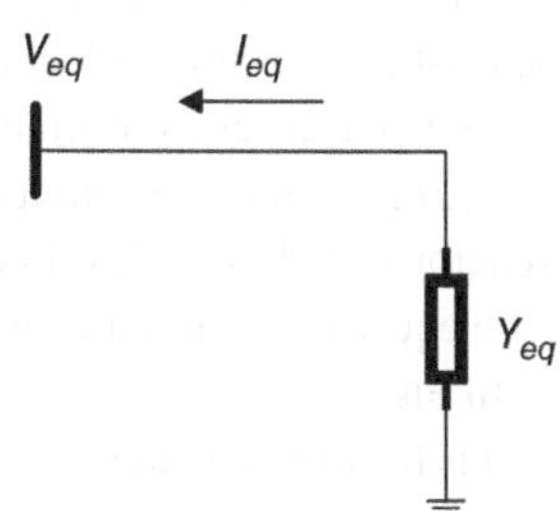

Figure 3.20 Shunt connected device.

Capacitor and SVC in I&C power systems are designed and used to provide passive reactive power into the network for reactive power compensation, power fact correction, and voltage support. SVC can also provide inductive power support when it is needed. These devices are in shunt connection to the network through terminal buses, as shown in Figure 3.20.

Load flow, motor starting, and transient stability analyses model. For these studies and analysis, the capacitor and var compensator can be modeled as constant impedance load with the equivalent shunt impedance calculated from the capacitance and var rating, i.e.

$$Z_{\text{eq}} = \frac{1}{Y_{\text{eq}}} = \frac{1}{\omega C} \text{ or } Z_{\text{eq}} = \frac{V^2}{\text{Var}} \tag{3.20}$$

where,

V and Var are the device-rated voltage and reactive power, respectively.

Short-circuit analysis model. Short-circuit analysis focuses on the equipment and devices that contribute to fault currents. Capacitor and Var compensator are not sustainable fault current sources; they either are neglected or modeled by Figure 3.20 and Equation (3.20).

Harmonic analysis model. Due to the frequency-dependent nature of their impedance, the capacitor and Var compensator are modeled differently in harmonic analysis. Figure 3.21 is the equivalent circuit model for them.

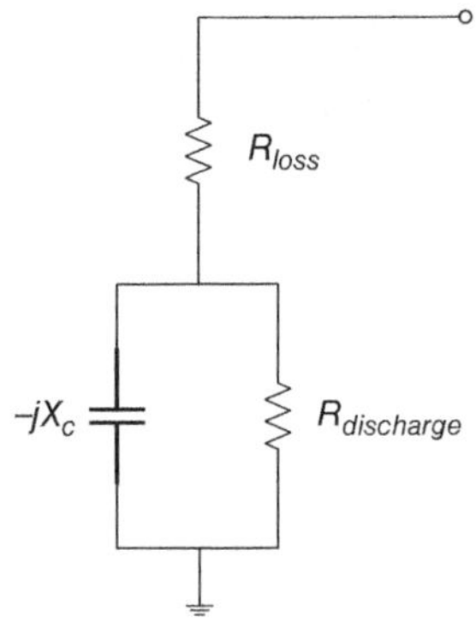

Figure 3.21 Capacitor model for harmonic analysis. Source: IEEE Std 3002.8-2018 [8]/with permission of IEEE.

Parameters in Figure 3.21 are:

h is harmonic order.

R_{loss} is the series resistance representing the equivalent loss (Ω).

$R_{discharge}$ is the parallel resistance representing the equivalent discharge resistance (Ω).

$X_c = \frac{1}{h\omega C}$ is capacitor impedance (Ω).

3.1.9 Short Line Reactor and Choker

Load flow, short-circuit, motor starting, and transient stability analyses model. Short-line reactor and choker are basically inductive elements; thus, they are modeled as a series branch without charging capacitance as in Figure 3.11.

Harmonic analysis model. Due to the frequency-dependent nature the short-line reactor and choker should be properly modeled for harmonic studies and analysis. The equivalent model is given in Figure 3.22.

In Figure 3.22,

h is harmonic order.

R is resistance.

X_h is reactance at h ($= h\omega L$).

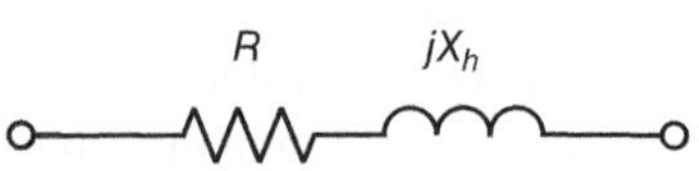

Figure 3.22 Reactor and choker model for harmonic analysis. Source: IEEE 399-1997 [12]/with permission of IEEE.

3.1.10 Harmonic Filter

Harmonic filters consist of a resistor, inductor, and capacitor. These are all passive components. Therefore, harmonic filters in power systems generally exhibit linear voltage versus current characteristics. Some commonly used harmonic filter structures are shown in Figure 3.23.

Load flow, short-circuit, motor starting, and transient stability analyses model. Harmonic filters are modeled as constant impedance in load flow, short-circuit, motor starting, and transient stability analyses. The impedance values are calculated based on the harmonic filter circuit structure and parameters at the fundamental frequency.

Harmonic analysis model. Harmonic filters are designed to provide lower impedances at turned frequencies, as can be seen from the notches in the impedance Z versus frequency f plots for each filter. In other words, their impedance varies as applied frequency changes. The harmonic filters introduced in Figure 3.23 serve different purposes. A single-tuned harmonic filter is used to suppress a specific harmonic at or near the tuned frequency. The second-order harmonic filter is often used to suppress higher frequencies. The third-order harmonic filter has one additional tuned frequency point compared to the second-order harmonic filter. A C-type harmonic filter has smaller losses at the fundamental frequency. Harmonic analysis will use the frequency-dependent impedance curves associated with each type of filter in modeling and simulation. All capacitance and inductance in the filters are the function of harmonic orders.

3.1.11 Harmonic Source

Harmonics are sinusoidal voltages or currents that have frequencies that are integer multipliers of the fundamental frequency. Common harmonic sources include power electronic devices and nonlinear loads such as variable frequency drives (VFDs) or adjustable speed drives (ASDs), uninterruptible power supplies (UPS), rectifiers, electronic ballasts, power converters, computers, IT equipment, etc. Harmonic sources can be in the format of voltage or current, depending on whether the device control is voltage-based or current-based. Harmonic sources in I&C power systems create adverse effects on the quality of electrical power delivery and consumption. The modeling process of harmonic sources allows computer simulation software tools to correctly replicate interactions from power electronic and nonlinear devices and their impacts on power systems.

Load Flow, Short-Circuit, Motor Starting, and Transient Stability Analyses Model. Harmonic sources are normally modeled as loads of various types according to their characteristics at the fundamental frequency. For example, in load flow analysis, the entire VFD drive train, including the drive, motor, and load, is modeled as a constant power load. Meanwhile, in short-circuit analysis, depending on

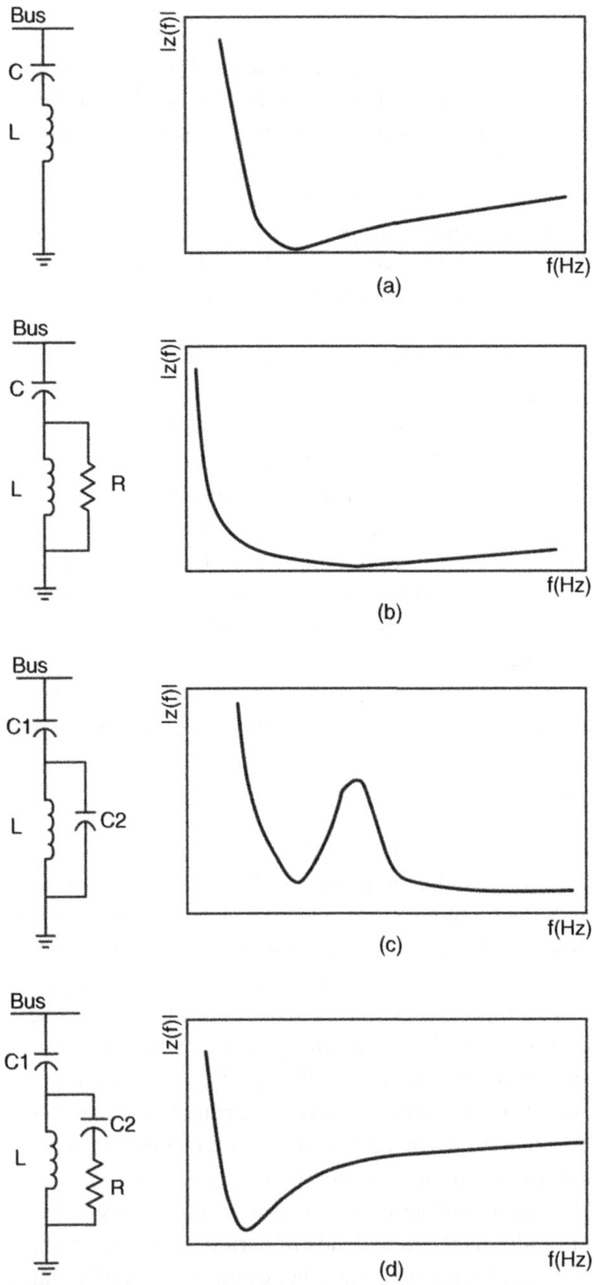

Figure 3.23 Harmonic filter model (a) Single-tuned filter, (b) High-pass filter (second order), (c) Undamped high-pass filter, and (d) High-pass C-type filter (third order). Source: IEEE Std 3002.8-2018 [8]/with permission of IEEE.

back-feeding current capability, a VFD can be modeled as a fault current source with limited capacity or is not modeled if it doesn't have back-feeding current capability. Furthermore, in motor starting and transient stability analyses, VFD controls should be modeled, which is discussed under the Section 3.1.12.

Harmonic analysis model. Harmonic sources are modeled by a spectrum, including the harmonic order and corresponding harmonic magnitudes and phase angles. The phase angle refers to the relative angle of a particular harmonic content with respect to the fundamental voltage or current phase angle. It is vital information to obtain accurate study results. Table 3.10 shows a sample harmonic current spectrum from a six-pulse motor drive at 480 Volt.

Harmonic source information can also be represented by actual voltage or current waveforms and harmonic content charts. Figure 3.24 shows a sample harmonic current waveform in a few cycles and the corresponding harmonic content chart for magnitudes. The harmonic spectrum can be obtained by applying Fourier Transformation to the waveform. Data in Figure 3.24 are not related to Table 3.10.

Some devices and equipment in power systems can generate voltages and/or currents containing components that are noninteger multipliers to the fundamental frequency. These components are called interharmonics. Devices and situations that can result in interharmonics in I&C power systems are cycloconverters, ASDs/VFDs, and loads that do not pulsate synchronously with the

Table 3.10 Sample harmonic current source model.

Harmonic order	Magnitude (%)	Relative angle (degree)
1	100.00	0.00
5	18.24	−55.68
7	11.90	−84.11
11	5.73	−143.56
13	4.01	−175.58
17	1.93	111.39
19	1.39	68.30
23	0.94	−24.61
25	0.86	−67.64
29	0.71	−145.46
31	0.62	176.83
35	0.44	97.40
37	0.38	54.36

Source: IEEE Std 3002.8-2018 [8]/with permission of IEEE.

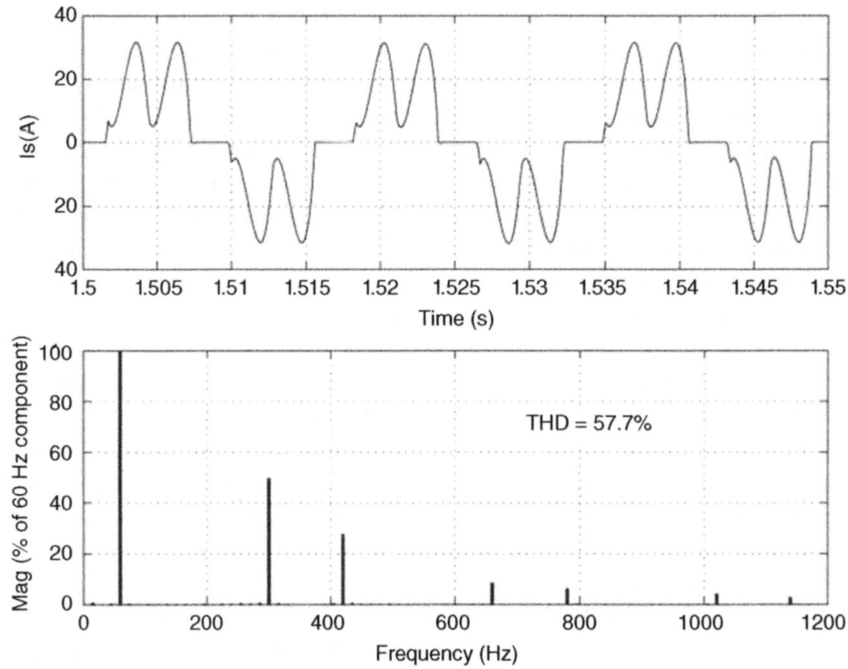

Figure 3.24 Sample harmonic waveform and content chart. Source: Singh et al. [13]/ with permission of IEEE.

fundamental power system frequency. Interharmonic frequencies generated from ASDs/VFDs and converter devices can be expressed in Equation (3.21)[8]:

$$f_i = (mp_1 \mp 1)f_n \mp np_2 f_z \tag{3.21}$$

where,

f_n is the system's fundamental frequency.
f_z is the modulation frequency.
p_1 is the rectifier's number of the pulses.
p_2 is the inverter's number of pulses.
m takes 0, 1, 2, 3, …
n takes 1, 2, 3, …

For each interharmonic content, its magnitude and relative phase angle should be provided and modeled using Table 3.10 format.

3.1.12 Power Converter

Power converters are electronic devices that are used to convert AC power to DC power or vice versa. In I&C power systems, power converters are extensively used

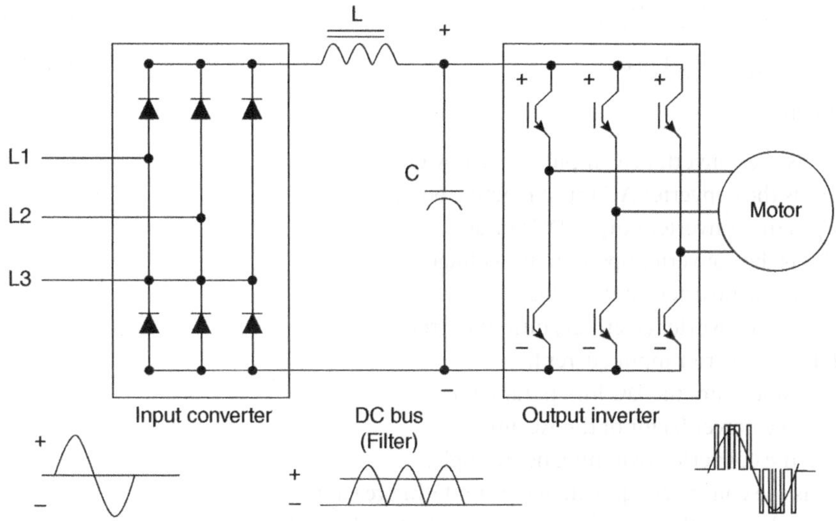

Figure 3.25 A Simplified Schematic of an ASD. Source: Dai et al. [14]/with permission of IEEE.

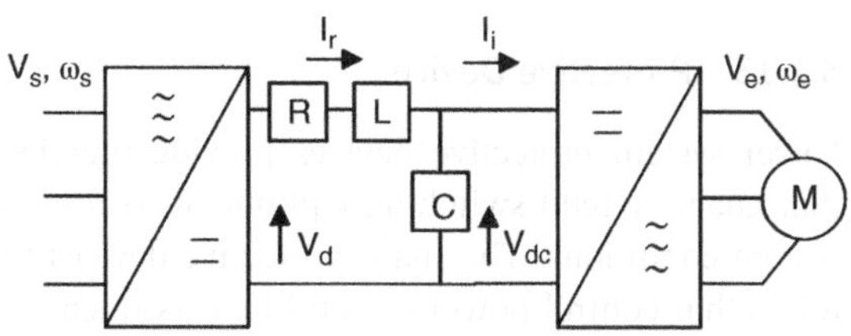

Figure 3.26 ASD Circuit Model Representation. Source: Dai et al. [14]/with permission of IEEE.

for motor drives, battery charging and discharging, and renewable power sources. Figure 3.25 is a simplified schematic for an ASD in motor speed and torque control. It is composed of an AC input converter, a DC link, and an AC output inverter. The equivalent circuit model for system analysis for Figure 3.25 is presented in Figure 3.26, which can further be described with a set of algebra equations with respect to the input side (AC line), including the DC link in Equation (3.22).

$$V_{d0} = \frac{3\sqrt{2}}{\pi} V_s \tag{3.22a}$$

$$R_s = \frac{3}{\pi} X_s \tag{3.22b}$$

$$\cos\varphi = \frac{1+\cos\mu}{2} \tag{3.22c}$$

$$V_d = V_{d0}\cos\varphi \tag{3.22d}$$

$$P_{ac} = P_{dc} = V_{do} I_r \tag{3.22e}$$

$$Q_{ac} = P_{dc} \tan \varphi \tag{3.22f}$$

where,

P_{ac} is the converter AC input active power.
Q_{ac} is the converter AC input reactive power.
V_d is the converter output DC voltage.
V_{d0} is the converter open-circuit voltage.
V_{dc} is the inverter input voltage.
I_r is the converter (rectifier) output current.
I_i is the inverter input current.
R, L, and C are the DC link parameters.
φ is the power factor of the AC inputs.
μ is the converter switching delay angle.
X_s is the converter equivalent commutating reactance.
V_s and ω_s are the AC side voltage and electrical speed.
V_e and ω_e are the motor terminal voltage and electrical speed, which is proportional to the frequency applied to the motor.

3.1.13 Protective Device

Power system protective devices provide intelligence and initiate the actions that enable circuit-switching equipment to respond to abnormal or dangerous system conditions. The main protective devices in I&C power systems include relays that control power circuit breakers rated above 600 V, current-responsive self-contained elements that operate multiple pole low-voltage circuit breakers and fuse alone or in combination with other suitable means that properly provide isolation of faulted or overloaded circuits. There are also special types of relays that respond to abnormal electric system conditions and may cause circuit breakers or other switching devices to disconnect defective equipment from the remainder of the system.

Overcurrent Relays. The most common relay for short-circuit protection of I&C power systems is the overcurrent relay, which is further classified into inverse type (A), very inverse type (B), extremely inverse type (C), and instantaneous type (D), each adjustable by their minimum and maximum time dial (TD) positions. A family of time current operating curves available with a typical inverse-time overcurrent relay is shown in Figure 3.27. The characteristics of different types of relays as a function of time dial are shown in Figure 3.28.

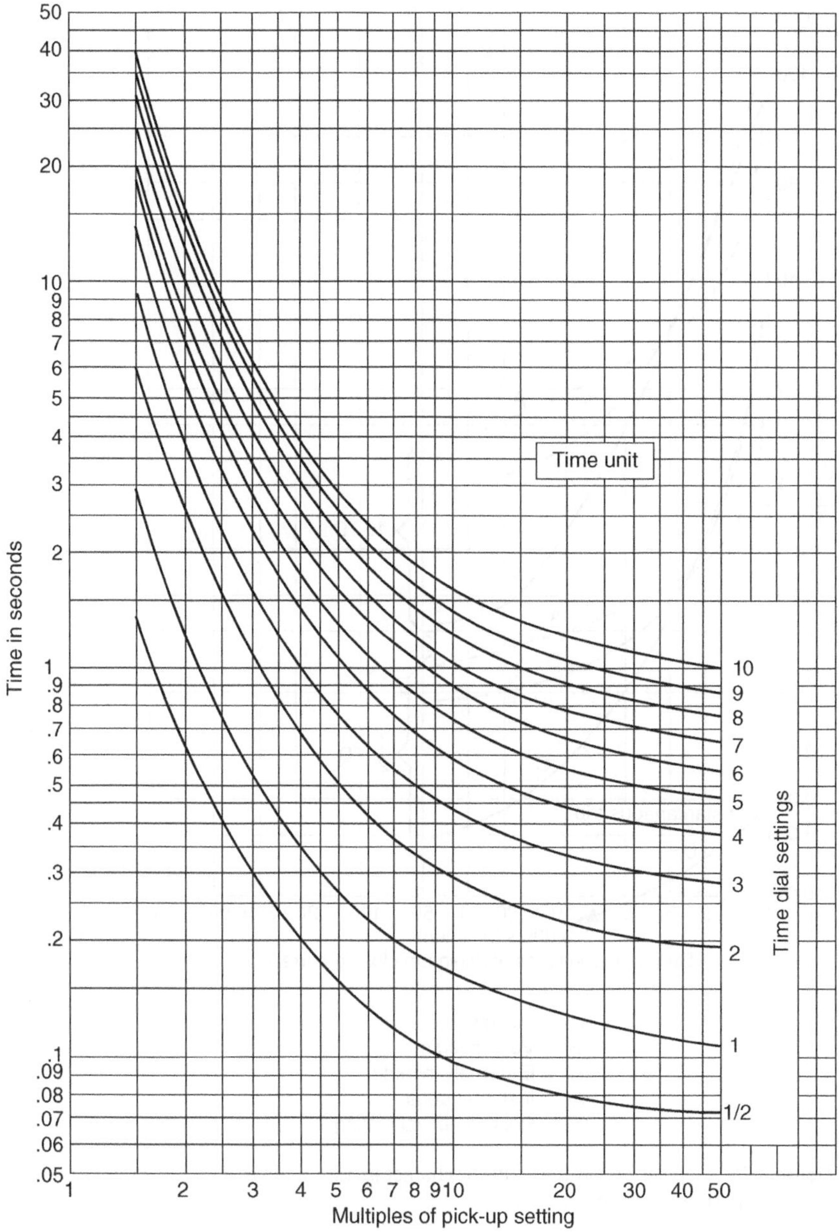

Figure 3.27 Time–current characteristics of a typical inverse-time overcurrent relay. Source: IEEE Std 141-1999 [15]/with permission of IEEE.

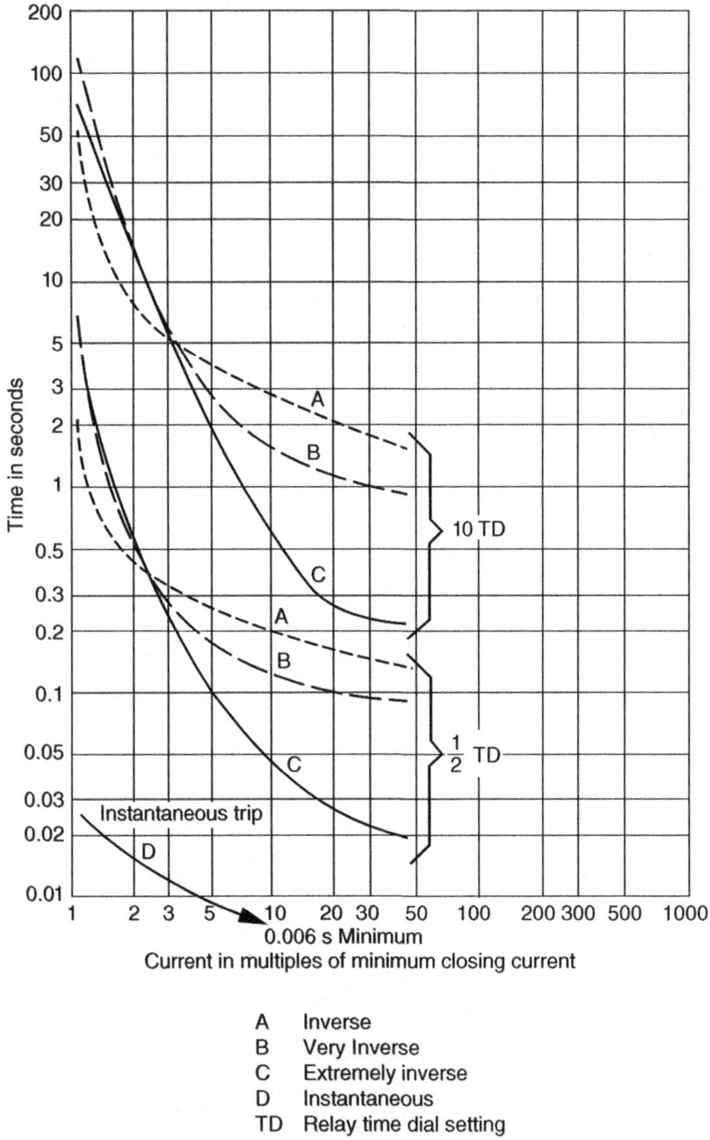

Figure 3.28 Typical relay time–current characteristics. Source: IEEE Std 141-1999 [15]/with permission of IEEE.

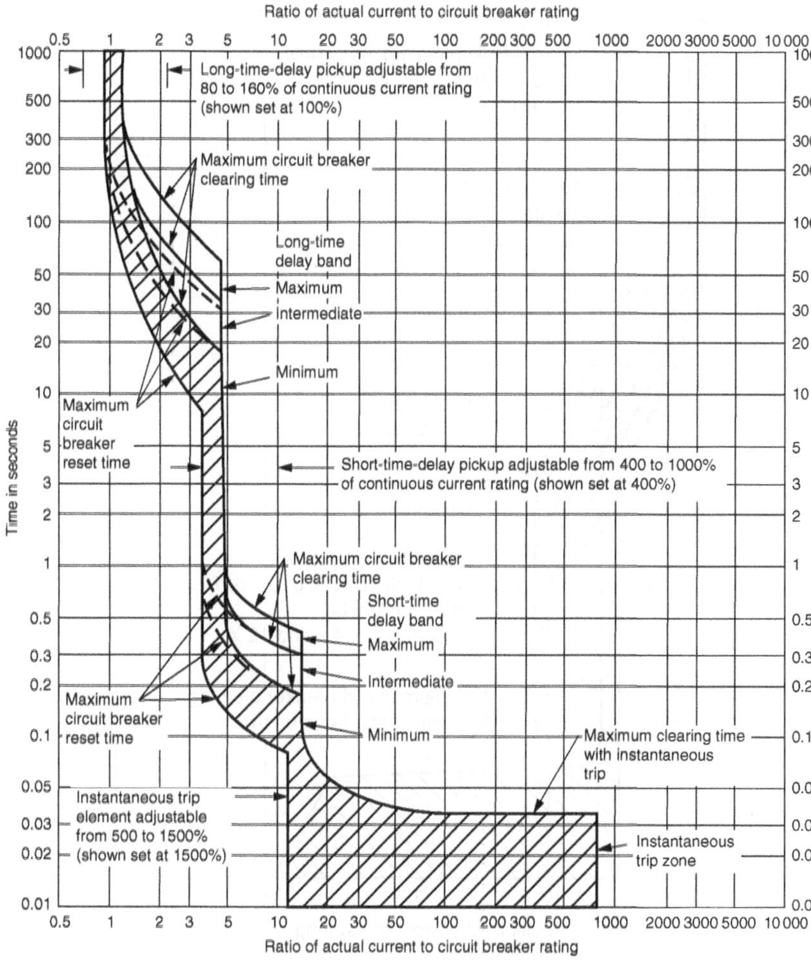

Figure 3.29 Typical time–current plot for electromechanical trip devices. Source: IEEE Std 141-1999 [15]/with permission of IEEE.

Trip Devices. There are two types of trip devices for low-voltage power circuit breakers. The electromechanical trip devices are characteristic curves with a very inverse shape and a broad tolerance band, as shown in Figure 3.29. Another type of trip devices are solid-trip devices whose typical time–current plots are shown in Figure 3.30.

Fuses. Fuses are an overcurrent protective device with a circuit opening fusible part that is heated and severed by the passage of overcurrent through it. Fuses suitable for the range of voltages in two general categories: power fuses (over 600 V) and low-voltage fuses (under 600 V). A typical time–current characteristic for

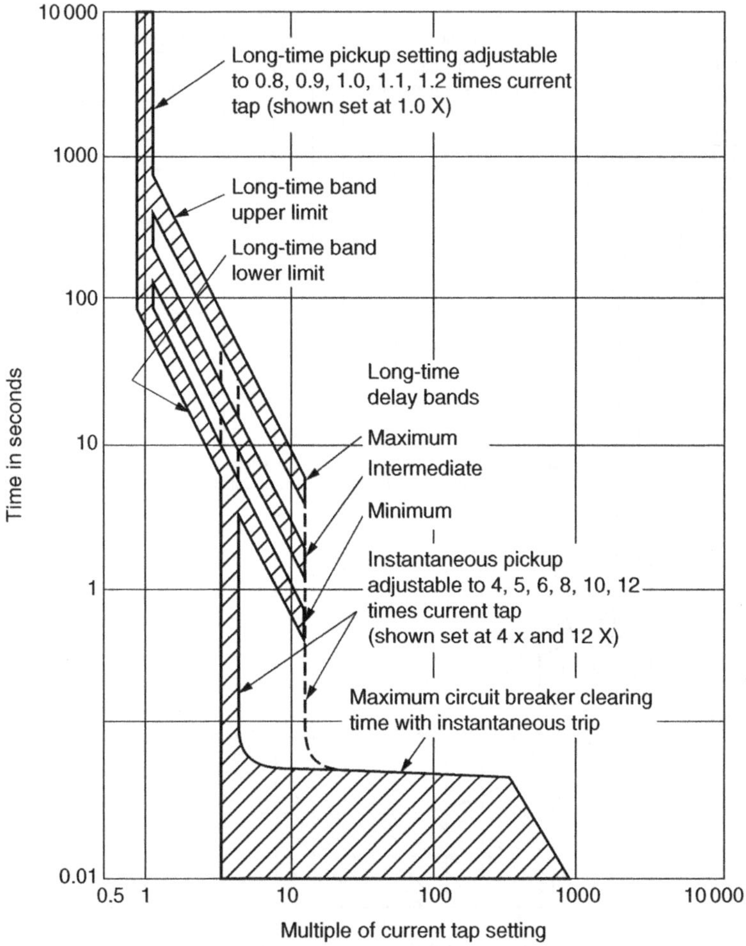

Figure 3.30 Typical time–current plot for solid-state trip devices. Source: IEEE Std 141-1999 [15]/with permission of IEEE.

a power fuse is shown in Figure 3.31. A typical time–current characteristic curve for electronic fuses is shown in Figure 3.32.

3.2 Control System Model

There exist many controls in power systems governing the behavior of equipment and devices to perform designed responses and provide protection. Such controls include a synchronous machine speed governor, excitation and AVR system,

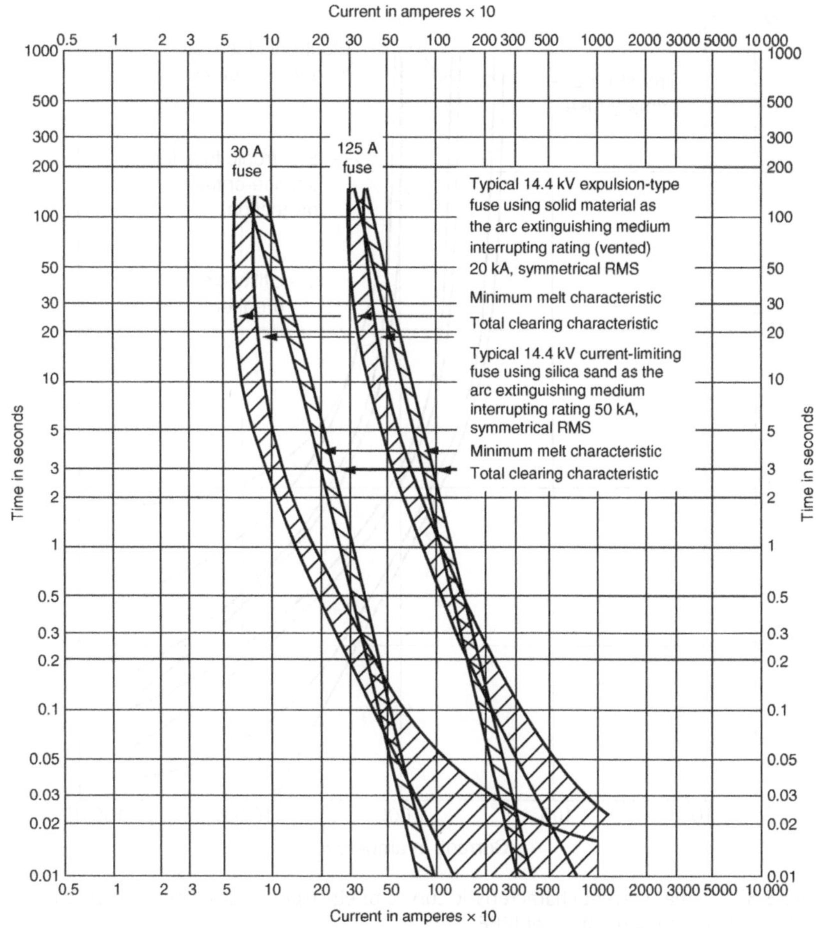

Figure 3.31 Time–current characteristic curves showing the difference between solid-material expulsion-type and current-limiting-type power fuses. Source: IEEE Std 141-1999 [15]/with permission of IEEE.

motor starting device, motor drive, reactive power compensator, transformer tap changer, capacitor switching, etc. Most of the control models involve differential equations with time constants, gains, and other parameters and are represented in transfer function block diagram format.

3.2.1 Generator Control

For transient stability analysis, the generator prime mover needs to be modeled since it is a key player in the generator shaft dynamics and makes a major

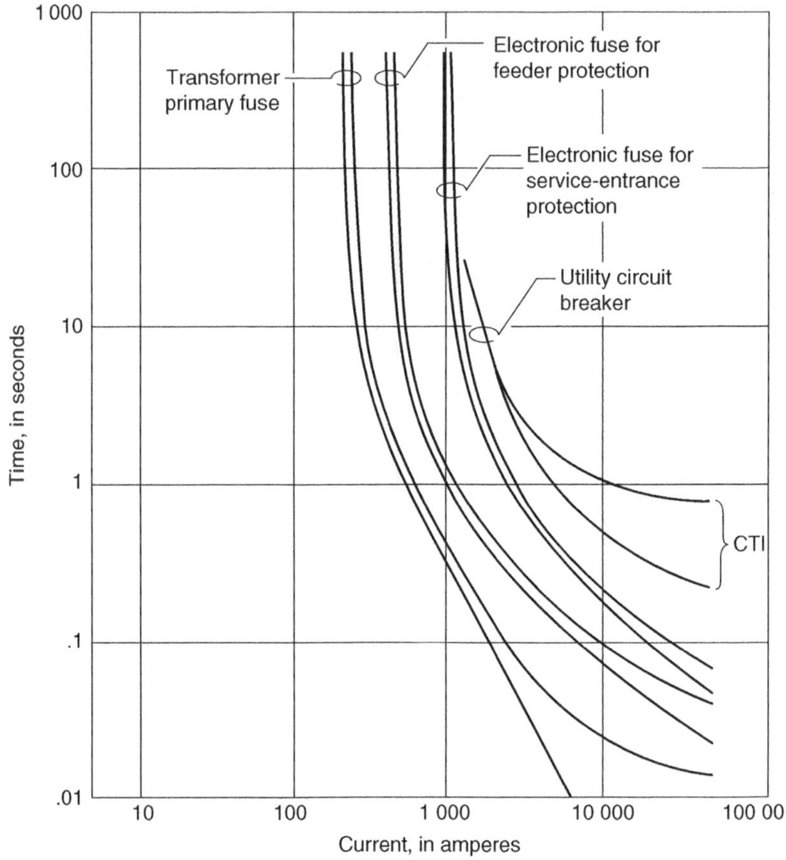

Figure 3.32 Time–current characteristic curves of electronic fuses. Source: IEEE Std 141-1999 [15]/with permission of IEEE.

contribution to the generator swing equation shown in Equation (3.3). The speed governor controls the interface with prime movers to govern generator speed and thus affect the system frequency. For synchronous machines, the excitation system needs to be modeled, too, as it is essential to provide a current source to filed windings for producing rotation magnetomotive force (MMF) to interact with stator MMF to convert shaft mechanical torque to electrical power at stator winding outputs. AVR is to regulate generator output voltage through its control function, and it is a part of the excitation system.

Prime Mover and Speed Governor Model. The governor controls the mechanical input from the prime mover to the generator to maintain the desired speed and power output. The combined prime mover and speed governor system is usually modeled together by transfer functions, which include the prime mover mechanics, speed sensing signal, output power setting, speed governor, and its feedback control loops. The typical model measures the speed and compares it with the speed reference to adjust power output through the governor droop setting. Figure 3.33 shows an IEEE steam turbine with a governor control model.

Prime mover and speed governor model parameters should be provided by vendors or identified by testing or simulations. Sample data with units for Figure 3.33 model is shown in Table 3.11.

The control model in Figure 3.33 is in a transfer function block diagram, which can be written into the ordinary different equation format for applying numerical solutions. For the following sample transfer function block diagram in Figure 3.34, a corresponding differential equation set can be derived in Equations (3.23), or in a compact form in Equations (3.24).

$$\frac{dx_1}{dt} = \frac{1}{T_1}(x_{\text{in}} - x_1) \tag{3.23a}$$

$$\frac{dx_2}{dt} = \frac{1}{T_2}(K_2 x_{\text{ref}} - K_2 x_1 - K_2 x_4 - x_2) \tag{3.23b}$$

$$\frac{dx_3}{dt} = \frac{1}{T_3}(x_2 - K_3 x_3) \tag{3.23c}$$

$$\frac{dx_4}{dt} = \frac{1}{T_4}\left(K_4 \frac{dx_3}{dt} - x_4\right) \tag{3.23d}$$

$$x_{\text{out}} = x_3 \tag{3.23e}$$

where,

x are state variables.
x_{in} is the input.
x_{out} is the output.
x_{ref} is the reference.
K_s and T_s are gains and times constants.

Organize them into state space equations in a compact form,

$$\dot{x} = Ax + Bu \tag{3.24a}$$

$$x_{\text{out}} = Cx \tag{3.24b}$$

where,

x is a variable vector representing the internal states of the speed governor and prime mover system.
u is the input signal normally consisting of speed input ω and speed reference ω_{ref}.
x_{out} is the output shaft mechanical power P_m.

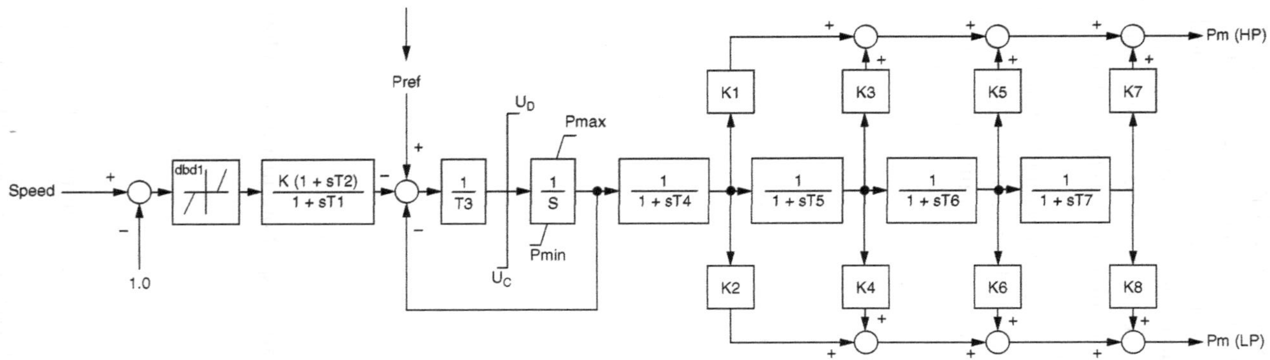

Figure 3.33 The IEESGO steam turbine and speed governor model. Source: IEEE [16]/with permission of IEEE.

Table 3.11 Sample data for IEESGO model.

Parameters	Description	Units
K	Governor gain (1/droop)	pu
T_1	Lag time constant	s
T_2	Lead time constant	s
T_3	Valve position time constant	s
U_0	Maximum valve opening rate	pu/s
U_c	Maximum valve closing rate	pu/s
P_{max}	Maximum valve opening on MW capability	pu
P_{min}	Minimum valve opening on MW capability	pu
P_{ref}	Output power reference	pu
P_m *(HP)*	Output power from the HP channel	pu
P_m *(LP)*	Output power from the LP channel	pu
T_4	Time constant for steam inlet	s
K_1	HP fraction	
K_2	LP fraction	
T_5	Time constant for second boiler pass	s
K_3	HP fraction	
K_4	LP fraction	
T_6	Time constant for third boiler pass	s
K_5	HP fraction	
K_6	LP fraction	
T_7	Time constant for fourth boiler pass	s
K_7	HP fraction	
K_8	LP fraction	
db1	Deadband	
Speed	Machine speed	pu

Source: IEEE [16]/with permission of IEEE

A, B, C are the system parameter matrices.

Excitation and AVR Model. Excitation and AVR work together to provide the field current of the generator and to regulate the generator terminal voltage. The model includes the excitation source, AVR, and any associated control loops. Figure 3.35 shows the IEEE Type DC2C excitation system model. And Table 3.12 is the model associated with sample data.

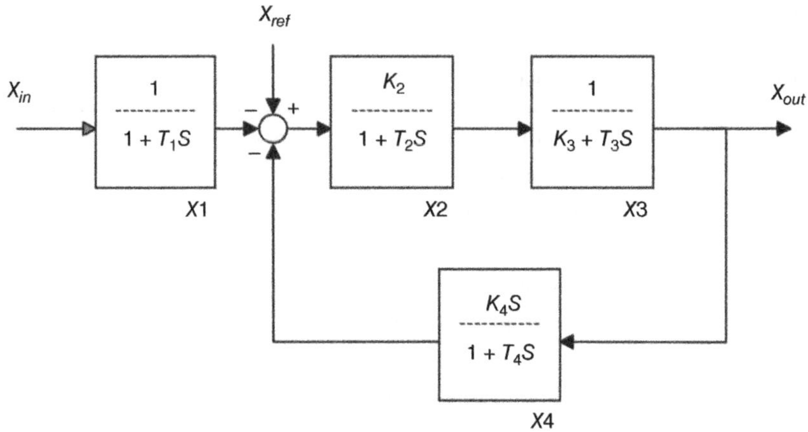

Figure 3.34 A sample transfer function block diagram.

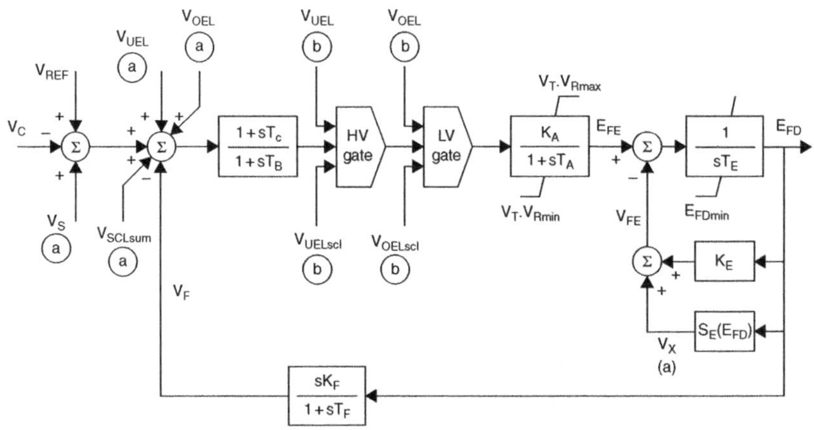

Figure 3.35 The IEEE Type DC2C excitation system model. Source: IEEE Std 421.5-2016 [17]/with permission of IEEE.

The typical inputs of the model are monitored voltage and voltage settings. The model output is the filed winding excitation voltage. The model transfer function block diagram Figure 3.35 shows many additional inputs from supplementary controls, but the main inputs are system voltage and reference voltage. It is worth noting that the system voltage can be at the terminal of the generator or a point a few transformations away. In the latter case, the voltage at a remote location can be controlled.

As in the speed governor and prime mover example, transfer function block diagrams for excitation and AVR systems can also be written into differential

Table 3.12 Sample data for IEEE Type DC2C excitation system model.

Symbol	Description	Value	Units
R_C	Resistive component of load compensation	0	pu
X_C	Reactance component of load compensation	0	pu
T_R	Regulator input filter time constant	0	s
K_A	Regulator output gain	300	pu
T_A	Regulator time constant	0.01	s
T_B	Regulator denominator (lag) time constant	0	s
T_C	Regulator numerator (lead) time constant	0	s
K_E	Exciter field proportional constant	0	pu
T_E	Exciter field time constant	1.33	s
VR_{max}	Maximum controller output	4.95	pu
VR_{min}	Minimum controller output	−4.95	pu
KF	Rate feedback gain	0.02	pu
TF	Rate feedback time constant	0.675	s
T_1	Exciter output voltage for saturation factor SE(E1)	3.05	pu
SE_1	Exciter saturation factor at exciter output voltage E1	0.279	
T_2	Exciter output voltage for saturation factor SE(E2)	2.29	pu
SE_2	Exciter saturation factor at exciter output voltage E2	0.117	

Source: IEEE Std 421.5-2016 [17]/with permission of IEEE.

equations in a compact state space format:

$$\dot{x} = Ax + Bu \tag{3.25a}$$

$$X_{\mathrm{out}} = Cx \tag{3.25b}$$

where,

x is the state variable vector in the excitation and AVR system.
u is the input signal normally consisting of voltage input V_t and voltage reference V_{ref}.
X_{out} is the output excitation voltage E_{fd}.
A, B, C are the system parameter matrices.

3.2.2 Motor Starting Control

During starting, magnetic flux inside the motor air gap is building up, which requires inductive current, or reactive power, supplied from the power source.

The typical starting current is six or more times the motor-rated current at rated voltage. Equivalently speaking, during the starting, the motor exhibits considerably smaller inductive impedance to the external system, which causes low-voltage conditions in the connected network. Various motoring starting control and devices are configured trying to mitigate the voltage drop issue. Some configurations attempt to reduce motor starting voltage, which in turn reduces the starting current. However, the tradeoff is that at the reduced voltage, motor torque will be reduced too. This means the motor will take a longer time to accelerate to its normal operating speed. If the voltage is reduced too much, the motor will not produce enough torque and in extreme cases it will not be successfully started. Several common motor starting configurations employed in I&C power systems are discussed below. Their models are provided to help perform simulation studies to verify motor starting scenarios in both the design and operation phases.

Direct On-line (DOL) Start. A DOL start is the simplest, most common, and least expensive method of starting motors. DOL start offers high-acceleration torque and reduced acceleration time. On the other hand, DOL start will draw the maximum motor current during start, which can cause voltage drops to cross the network. The mechanical torque produced by DOL start is computed from the following equation based on a single-cage motor equivalent circuit model:

$$T_d = 3\frac{R_r}{s\omega_r} \times \frac{|V_{\text{ph}}|^2}{\sqrt{\left(R_{e1} + \frac{R_r}{s}\right)^2 + (X_{e1} + X_r)^2}} \tag{3.26}$$

where,

T_d is the motor shaft torque.
R_r and X_r are the motor equivalent circuit model stator resistance and reactance.
R_{e1} and X_{e1} are the motor rotor resistance and reactance referred to the stator side.
s is the motor slip.
ω_r is the synchronous speed.
V_{ph} is the motor phase voltage.

It can be seen from Equation (3.26) that the torque developed by the motor at any speed is approximately proportional to the voltage squared, as illustrated in Figure 3.36.

In contrast with the full-voltage starting, several different methods of reduced voltage are starting to become standard practice in I&C industrial power systems.[7]

Wye/delta (Y-Δ) start. For small- and medium-sized motors, the wye/delta starter can be used. The stator winding is connected in wye during the starting and in delta for the running, as shown in Figure 3.37.

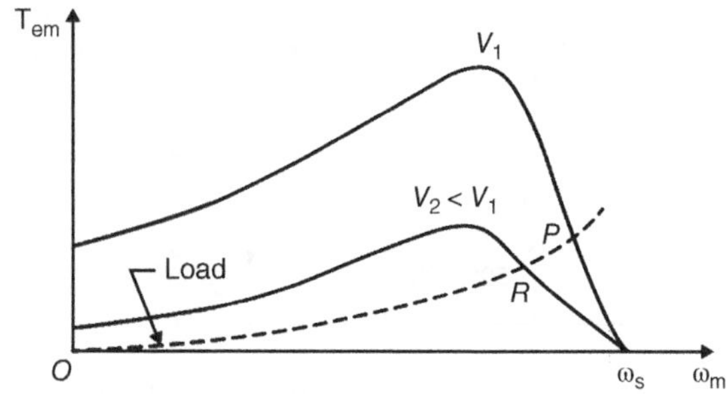

Figure 3.36 Effect of voltage variation to motor torque. Source: IEEE Std 421.5-2016 [17]/with permission of IEEE.

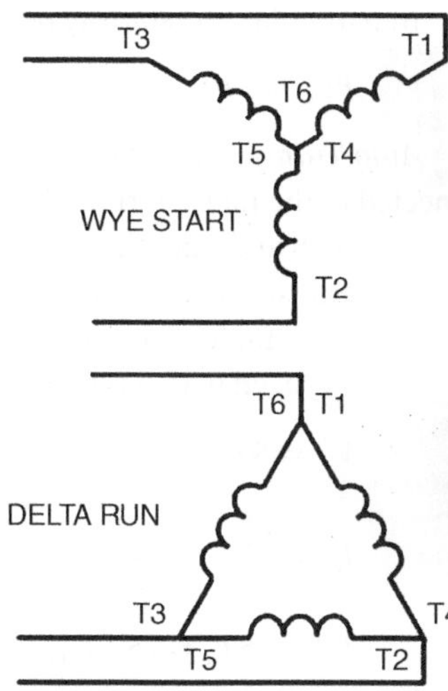

Figure 3.37 Y-Δ start configuration.

Since during the start, voltage is reduced by a square root of 3, therefore, the starting current is reduced to 58% of the DOL start, and starting torque is reduced to 33% of the DOL start, respectively:

$$V_{\text{st}} = \frac{V_{\text{DOL}}}{\sqrt{3}} = 0.58V_{\text{DOL}} = 58\%V_{\text{DOL}} \tag{3.27a}$$

$$I_{\text{st}} = \frac{V_{\text{st}}}{Z_{lr}} = 0.58\frac{V_{\text{DOL}}}{Z_{lr}} = 58\%I_{\text{DOL}} \tag{3.27b}$$

$$T_{\text{st}} \propto V_{\text{st}}^2 = (0.58V_{\text{DOL}})^2 = 0.33V_{\text{DOL}}^2 \propto 33\%T_{DOL} \tag{3.27c}$$

where,

I_{st}, V_{st}, T_{st} are the motor starting current, terminal voltage, and starting torque with a Y-Δ start configuration.

I_{DOL}, V_{DOL}, T_{DOL} are motor starting current, terminal voltage, and starting torque with DOL start configuration.

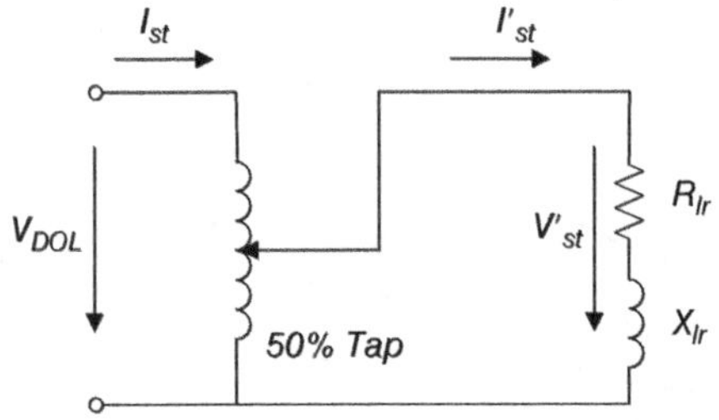

Figure 3.38 Autotransformer start equivalent circuit.

Z_{lr} is the equivalent motor starting impedance (locked-rotor impedance).

In particular, the starting power factor remains the same, i.e.

$$PF_{st} = PF_{DOL} \tag{3.27d}$$

Auto-Transformer Start. By setting the tap of an autotransformer that is connected to the motor terminal, the motor voltage can be reduced by the tap setting. Figure 3.38 and Equation (3.28) show the configuration and voltage reduction.

Autotransformer motor starters are usually provided with 80%, 65%, and 50% taps. For example, if the tap is set to 50%, then starting voltage, current, and torque will be reduced with a reference to the DOL start to:

$$V'_{st} = 50\%\ V_{DOL} \tag{3.28a}$$

$$I'_{st} = \frac{V'_{st}}{Z_{lr}} = \frac{0.5\ V_{DOL}}{Z_{lr}} = 50\%\ I_{DOL} \tag{3.28b}$$

$$I_{st} = 0.5\ I'_{st} = 25\%\ I_{DOL} \tag{3.28c}$$

$$T_{st} \propto V'^2_{st} = 0.25\ V^2_{DOL} \propto 25\%\ T_{DOL} \tag{3.28d}$$

The starting power factor remains the same, i.e.

$$PF_{st} = PF_{DOL} \tag{3.28e}$$

In-line Resistor Start. Inserting a resistor in-line (serial connection) with the motor can also reduce the motor terminal voltage to assist the starting, as shown in Figure 3.39 and Equation (3.29).

Motor starting voltage, current, torque, and power factor corresponding to 50% tap position are given below:

$$V'_{st} = 50\%\ V_{DOL} \tag{3.29a}$$

$$I_{st} = \frac{V'_{st}}{Z_{lr}} = \frac{0.5\ V_{DOL}}{Z_{lr}} = 50\%\ I_{DOL} \tag{3.29b}$$

$$T_{st} \propto V'^2_{st} = 0.25\ V^2_{DOL} \propto 25\%\ T_{DOL} \tag{3.29c}$$

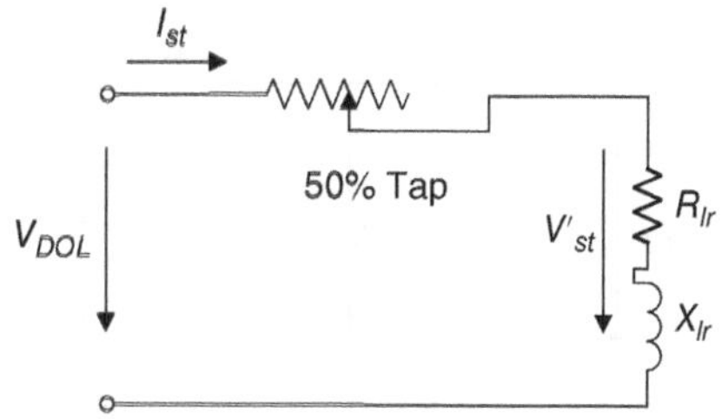

Figure 3.39 In-line resistor start equivalent circuit.

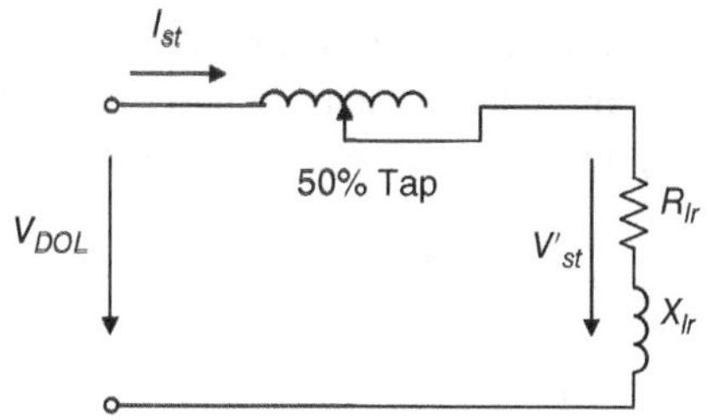

Figure 3.40 In-line reactor start equivalent circuit.

A new starting power factor shall be computed in this case due to a resistance change in the loop. It can be derived that the starting power factor be,

$$PF_{\text{st}} = \sqrt{1 - 0.5^2\left(1 - PF_{\text{DOL}}{}^2\right)} = \sqrt{1 - 0.25\left(1 - PF_{\text{DOL}}{}^2\right)} \tag{3.29d}$$

In-line Reactor Start. Insert a reactor in-line in Figure 3.40 with the motor work in a same way as a resistor, with only the change in effective power factor of the equivalent circuit.

For a tap position of 50%, motor starting voltage, current, torque and power factor are given below:

$$V'_{\text{st}} = 50\%\ V_{\text{DOL}} \tag{3.30a}$$

$$I_{\text{st}} = \frac{V'_{\text{st}}}{Z_{\text{lr}}} = \frac{0.5\ V_{\text{DOL}}}{Z_{\text{lr}}} = 50\%\ I_{\text{DOL}} \tag{3.30b}$$

$$T_{\text{st}} \propto V'^2_{\text{st}} = 0.25\ V^2_{\text{DOL}} \propto 25\%\ T_{\text{DOL}} \tag{3.30c}$$

The starting power factor will change in this case too due to be added reactance in the loop. It can be derived that the starting power factor at 50% tap will be,

$$PF_{\text{st}} = 50\%\ PF_{\text{DOL}} \tag{3.30d}$$

Shunt Capacitor. A capacitor can be switched on with a motor starting, which compensates for part of the reactive kvar drawn by the starting motor, as illustrated in Figure 3.41. The capacitor size is usually selected to provide about 50% of the motor-starting kvar. The capacitor is switched off at about 95% of the motor-rated speed, and the motor current drops to about full load.

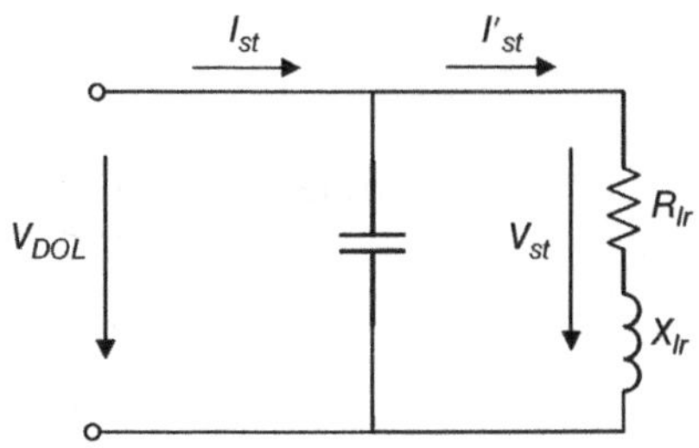

Figure 3.41 Shunt capacitor start equivalent circuit.

The starting voltage and starting torque will remain the same as DOL, but on the line side, the starting current will be reduced, and the power factor will be improved. Take an example of 50% kvar compensation, results are:

$$V_{\mathrm{st}} = V_{\mathrm{DOL}} \tag{3.31a}$$

$$I_{\mathrm{st}} = 50\%\ I'_{\mathrm{st}} \tag{3.31b}$$

$$T_{\mathrm{st}} = T_{\mathrm{DOL}} \tag{3.31c}$$

$$PF_{\mathrm{st}} = \frac{\sqrt{P^2 + Q^2}}{\sqrt{P^2 + 0.25Q^2}} PF_{\mathrm{DOL}} \quad \text{or} \quad PF_{\mathrm{st}} > PF_{\mathrm{DOL}} \tag{3.31d}$$

More starting method options can be found in the standard, including: [7]

1) Partial winding
2) Softer starter
3) Captive transformer
4) Electronic soft-starters
5) VFD/ASD

3.2.3 Adjustable Speed Control

Per IEEE Std. 100 [18] definition, "ASD and VFD is an electric drive designed to provide an easily operable means of speed adjustment of the motor, within a specified speed range." The definition also refers to the motor as being part of the adjustable speed system, implying that it is a system rather than a single device. ASDs are normally equipped with controllers to control the voltage and electrical speed (or frequency) supplied to the motor. By means of frequency regulation, controller parameters are set to achieve various designed objectives, typically a constant motor torque, a higher efficiency, or other desired motor dynamic performance. Switching time and magnitudes between different stages of control are also adjustable. Figure 3.42 shows a generic ASD speed controller to control ASD output voltage and frequency.

This controller consists of a proportional and integral (PI) block for speed control, a volts/Hz regulator, and a speed sensor block. The gains and time constants reflect the speed sensor dynamics, inverter characteristics, and the embedded control algorithm. The controller takes the motor's electrical speed as feedback, compares it with the reference speed to produce a speed error signal. The reference speed can be a preset value, or it can be adjusted during motor operation. When the reference speed is adjusted to a new setting, a speed error signal will be introduced. In this case, from the output of the PI control block, the electrical speed is changed to track the new reference speed, and the volts/Hz regulation is to maintain the proper V_e/ω_e ratio. Figure 3.42 parameter definitions are listed in Table 3.13.

3.2.4 Transformer Control

The transformer has the ability to regulate voltage under varying loads and conditions based on the design and operation specifications. This is done by tap changers to control the transformer's turn ratio and regulate the output voltage. Tap changers can be at fixed taps or set to online load tap changers (OLTC). In the case of OLTC, it is important to know where the voltage measurements are taken and what are the voltage set points. Computer models of transformer tap control allow simulation studies in load flow, short-circuit, motor starting, and transient

Figure 3.42 A Generic ASD Speed Controller. Source: Dai et al. [14]/with permission of IEEE.

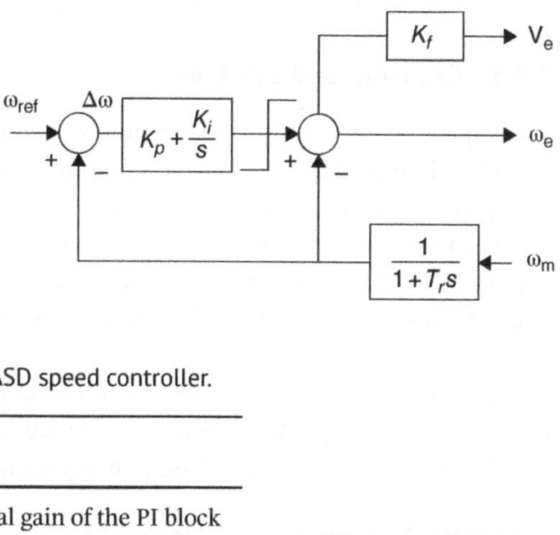

Table 3.13 Parameters of generic ASD speed controller.

Symbol	Description
K_p	Proportional gain of the PI block
K_i	Integral gain of the PI block
T_r	Time constant for the sensor block
K_f	Voltage gain volts/Hz regulator
ω_m	Motor electrical speed
ω_{ref}	Reference speed
V_e	Inverter output voltage
ω_e	Electrical speed output

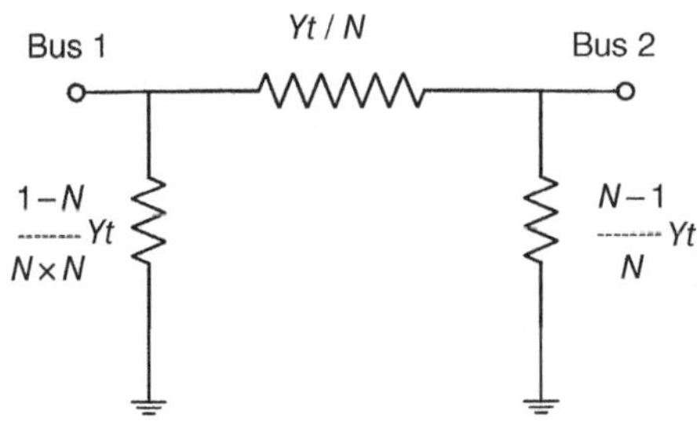

Figure 3.43 Tap changer model for two-winding transformer.

stability analyses to investigate the impact of transformer voltage regulations under various operating conditions and events. The transformer tap changer is modeled by adjusting transformer impedance as depicted in Figure 3.43 in a two-winding transformer example.

In Figure 3.43,

Z_t = is transformer winding impedance.
$Y_t = \frac{1}{Z_t}$ is transformer admittance.
$N = 1 + \frac{n}{100}$ is transformer turn ratio.
$n = 100(N - 1)$ is tap setting.

3.2.5 Capacitor and SVC Control

The capacitor and SVC inject or supply reactive power to the network. Shunt capacitor banks and SVC can be configured to be automatically controlled in I&C power systems to improve power factor and regulate voltage, as well as to reduce losses in the network. Computer models of these controls allow simulation studies in load flow, short-circuit, motor starting, and transient stability studies and analyses to investigate the behavior of these controls under various operating conditions.

The controls for capacitor banks are done manually or automatically by connecting or disconnecting segments of capacitance to the bus, as shown in Figure 3.44 to adjust shunt admittance Y based on Equation (3.32).

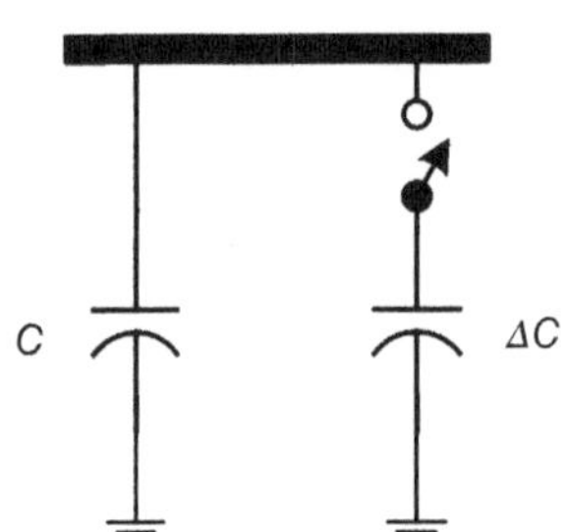

Figure 3.44 Switched capacitor bank model.

$$Y_{old} = j\omega C \text{ is the existing shunt admittance} \tag{3.32a}$$

$$\Delta Y = j\omega \Delta C \text{ is the incremental shunt admittance} \tag{3.32b}$$

$$Y_{new} = Y_{old} + \Delta Y \text{ is the modified shunt admittance} \tag{3.32c}$$

For SVC devices, the controls are adjusted to inject capacitive or inductive current into the network based on the voltage settings and feedback at the monitored bus. Current injection is achieved by adjusting the equivalent susceptance of the SVC. Figure 3.45 shows the structure of SVC, which consists of fixed capacitors and controlled inductors. A basic regulator transfer function diagram in Figure 3.46 to control bus voltage by producing required susceptance command based on the measured voltage errors at the control bus.

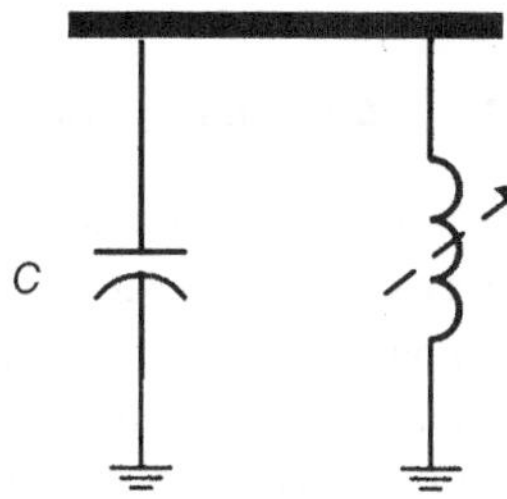

Figure 3.45 A generic SVC model.

The control model takes voltage error signal by comparing the actual voltage and the reference voltage then produce a new compensation susceptance reference B_{ref} for the SVC model by noticing that,

$$B = B_C + B_L = \omega C - \frac{1}{\omega L} \tag{3.33a}$$

$$Y = jB \tag{3.33b}$$

3.3 Network Modeling

Electrical networks in I&C power systems are modeled in algebra equations in phasor format to represent electrical components, including sources, loads, series and shunt equipment, and their interconnections within the network. These models are essential to define the system information so that the unknown variables, typically bus voltages, branch currents, source, and load powers, can be deterministically and uniquely solved. The common representations of the algebra equations include Y-Bus modeling and Z-Bus modeling, both sufficiently providing system topological and electrical information that are required to solve

Figure 3.46 A basic SVC voltage regulator model. Source: IEEE Special Stability Controls Working Group [19]/with permission of IEEE.

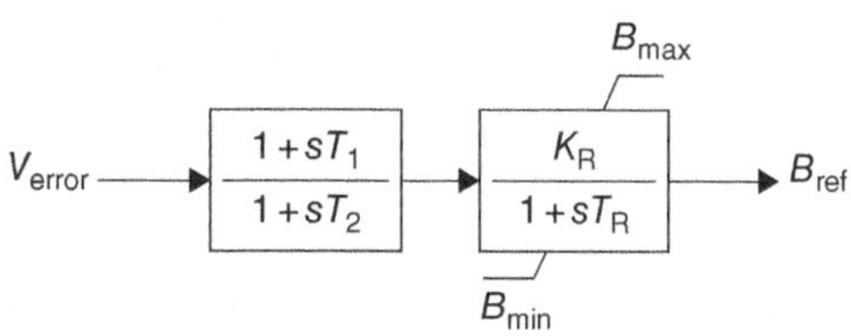

voltages and currents and associated variables in the system. In this section, the formulation process of the admittance equation and impedance equation based on the knowledge of the previously covered context and the network topology will be discussed. Special handling of sources and loads that exhibit a nature as constant voltage, current, or power is also addressed.

3.3.1 Admittance Equation and *Y*-Bus

The admittance equation is a fundamental approach in computer modeling of I&C power systems. It is a mathematical representation of the electrical admittance of the power system network, containing the complete network topology for electrical component interconnections and electrical parameters. In matrix form, the admittance equitation equation can be expressed as:

$$I_{\text{bus}} = Y_{\text{bus}} V_{\text{bus}} \tag{3.34}$$

where,

$Y_{\text{bus}} = Y + B$ is Y-Bus matrix.

The Y-Bus matrix represents the admittance values of the system network, including nodal admittance, which are contributed from nodal connected components Y matrix (sources, loads, shunt devices, and Pi-equivalent circuits), and from branch connected components B matrix (transformers, lines, cables, etc.).

I_{bus} is nodal (bus) current injection vector representing all shunt-connected equipment (sources, loads, shunt devices, etc.)

V_{bus} is nodal (bus) voltage vector.

Taking a simple three-bus example in Figure 3.47, by applying basic Kirchhoff's laws, the admittance equation for this system can be written in Equation (3.35).

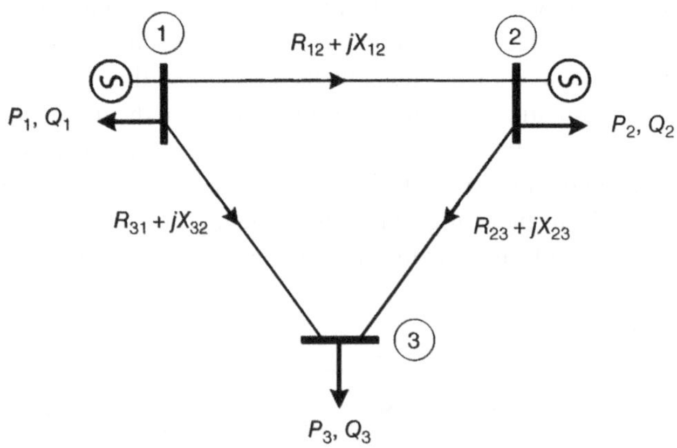

Figure 3.47 A simple three-bus system example.

$$I_1 = (y_{21} + y_{31})V_1 - y_{12}V_2 - y_{13}V_3$$

$$I_2 = -y_{21}V_1 + (y_{12} + y_{23})V_2 - y_{23}V_3 \quad (3.35)$$

$$I_3 = -y_{31}V_1 - y_{32}V_2 + (y_{13} + y_{23})V_3$$

where,

$y_{ij} = y_{ji} = \frac{1}{Z_{ij}} = \frac{1}{R_{ij}+jX_{ij}}$ is the admittance of branch circuit i to j.

Thus, the admittance equation of Equation (3.35) can be expressed as:

$$\begin{bmatrix} I_1 \\ I_2 \\ I_3 \end{bmatrix} = \begin{bmatrix} (y_{21}+y_{31}) & -y_{12} & -y_{13} \\ -y_{21} & (y_{12}+y_{23}) & -y_{23} \\ -y_{31} & -y_{32} & (y_{13}+y_{23}) \end{bmatrix} \begin{bmatrix} V_1 \\ V_2 \\ V_3 \end{bmatrix} \quad (3.36)$$

which gives

$$Y_{\text{bus}} = \begin{bmatrix} (y_{21}+y_{31}) & -y_{12} & -y_{13} \\ -y_{21} & (y_{12}+y_{23}) & -y_{23} \\ -y_{31} & -y_{32} & (y_{13}+y_{23}) \end{bmatrix} = \begin{bmatrix} Y_{11}\, Y_{12}\, Y_{13} \\ Y_{21}\, Y_{22}\, Y_{23} \\ Y_{31}\, Y_{32}\, Y_{33} \end{bmatrix} \quad (3.37)$$

In summary, the diagonal and off-diagonal elements of Y_{bus} are defined by:

$$Y_{ii} = \sum_{j=1, j\neq i}^{N} y_{ij} + y_i$$

$$Y_{ij} = -y_{ij}$$

where i, j is the node (bus) number, y_i is from a shunt element at node i if any, and N is the total number of nodes.

The Y_{bus} and the admittance equation are determined based on the characteristics of the system components, taking into account their connections and the configuration of the system. They provide the bases and are to be solved under specific operating conditions for load flow, motor starting, harmonics, and transient stability analyses.

If the nodal voltage vector V is known, Equation (3.36) can be solved directly. This requires the system to be linear, shunt and load components can be expressed in admittance format (constant impedance).

3.3.2 Impedance Equation and *Z*-Bus

The impedance equation and Z-bus are also effective in modeling power systems. The relationship between V_{bus} and I_{bus} is also given by Equation (3.38):

$$V_{\text{bus}} = Z_{\text{bus}} I_{bus} \quad (3.38)$$

Comparing Equation (3.34) with Equation (3.38), it can be seen that:

$$Z_{\text{bus}} = Y_{\text{bus}}^{-1} \tag{3.39}$$

So for the three-bus system in Figure 3.47, the impedance equation will take the form of:

$$\begin{bmatrix} V_1 \\ V_2 \\ V_3 \end{bmatrix} = \begin{bmatrix} Z_{11}\, Z_{11}\, Z_{11} \\ Z_{11}\, Z_{11}\, Z_{11} \\ Z_{11}\, Z_{11}\, Z_{11} \end{bmatrix} \begin{bmatrix} I_1 \\ I_2 \\ I_3 \end{bmatrix} \tag{3.40}$$

which gives

$$Z_{\text{bus}} = \begin{bmatrix} Z_{11}\, Z_{11}\, Z_{11} \\ Z_{11}\, Z_{11}\, Z_{11} \\ Z_{11}\, Z_{11}\, Z_{11} \end{bmatrix} \tag{3.41}$$

where V_{bus}, I_{bus} have the same definitions as before, while elements in Z_{bus} are not so rivals. It involves the original circuit branch and shunt brunch impedance data, system topology, element connections, and matrix operations.

Although Z_{bus} is able to be obtained through the process by applying Kirchhoff's laws, it is equivalent to the matrix inversion of Y_{bus} which is more straightforward to construct. There are computational effective and accurate numerical algorithms to achieve the matrix inversion step by step.

Same as Y_{bus} and the admittance equation, Z_{bus} and the impedance equation fully represent the system network, including both electrical parameters and topology information. It is specifically efficient in performing the short-circuit calculation and analysis. In that case, the diagonal elements provide essential data for calculating a bus fault current which will be discussed in the later chapter. When the current vector I_{bus} is known, the impedance equation can be solved directly. This requires all impedance to be linear, and shunt components and loads can be represented as linear and constant impedance (admittance).

Note that both the admittance equation and impedance equation are very handy to describe the system. This is based on the assumption that all components in the system are linear. For nonlinear shunt components and loads, special handling will be required, which are discussed below.

3.3.3 Constant Power Components

There are some shunt components or loads in I&C power systems that change relatively slowly in magnitude of active and reactive powers, with behavior like constant load within a certain range of operating conditions (mainly voltage variations). Such load cannot be directly represented in the admittance and impedance matrices. However, through some equivalency, constant power load can be approximated as constant impedance load. Assuming one constant power load is defined as $P+jQ$, where P and Q are the active and reactive power, respectively, and hold near as constants. By applying the following conversion around an estimated voltage point $\widehat{V}$, the parallel circuit equivalent of the load can be obtained,

$$R_{\text{eq}} = \frac{\widehat{V}^2}{P} \tag{3.42a}$$

$$X_{\text{eq}} = \frac{\widehat{V}^2}{Q} \tag{3.42b}$$

or

$$y_{\text{eq}} = \frac{1}{R_{\text{eq}}} - j\frac{1}{X_{\text{eq}}} \tag{3.42c}$$

where R_{eq} and X_{eq} are corresponding parallel connected resistance and reactance, as shown in Figure 3.48, and y_{eq} is the equivalent of admittance connected to a bus which will contribute to Y_{bus}, i.e. the admittance matrix as a shunt element.

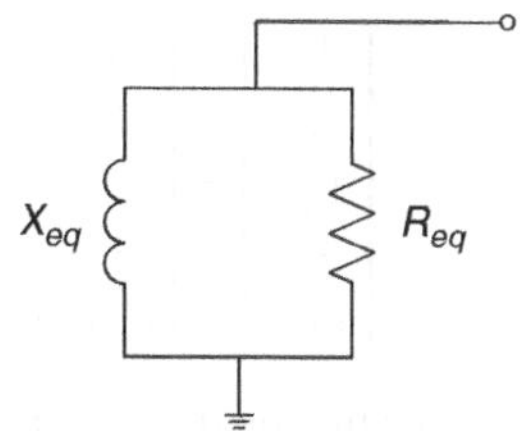

Figure 3.48 Constant power load model.

In computer simulation, constant power load or source model has certain limitations. Assuming a constant power load $\overline{S}_k = \overline{P}_k + j\overline{Q}_k$ connected to bus k, and the bus voltage becomes zero due to a fault. The load current would become to infinity if still let the load power is constant:

$$I_k = \frac{\overline{S}_k}{V_k} \to \infty (V_k \to 0) \tag{3.43}$$

To prevent such an ill condition in the simulation, there must be a bound $V_{\max}$ for the constant power assumption, for example:

$$S_k = \overline{S}_k, if\, V_k \geq V_{k,\max} \tag{3.44}$$

otherwise S_k will be represented as a constant impedance load,

$$R_{\text{eq}} = \frac{V_{k,\text{nom}}^2}{\overline{P}_k} \tag{3.45a}$$

$$X_{\text{eq}} = \frac{V_{k,\text{nom}}^2}{\overline{Q}_k} \tag{3.45b}$$

where $V_{k,\text{nom}}$ stands for bus nominal voltage, and R_{eq} and X_{eq} are parallel connected equivalent impedance at bus k.

3.3.4 Constant Voltage Components

Constant voltage components are mainly utility, generator, some motor drives.

Constant voltage in I&C power systems refers to a characteristic by design or nature of some components to maintain a stable voltage output despite variations in load conditions or currents. These types of components mainly include utility, generator with AVR function, UPS, transformer tap change-controlled bus, capacitor and AVR controlled bus, etc. They all interact with the network through terminal connected or controlled buses. Constant voltage buses are readily handled in the admittance equation since the assumption is the voltage vector V is known. Constant voltage sources are also readily handled in the impedance equation by eliminating the corresponding equations as demonstrated below. Assume the voltage $\overline{V}_k$ at bus k is fixed by a generator or utility source, then the corresponding impedance equation becomes:

$$\begin{bmatrix} V_1 \\ \vdots \\ \overline{V}_k \\ \vdots \\ V_N \end{bmatrix} = \begin{bmatrix} Z_{11} & \cdots & Z_{1k} & \cdots & Z_{1N} \\ \vdots & \ddots & \vdots & \ddots & \vdots \\ Z_{k1} & \cdots & Z_{kk} & \cdots & Z_{kN} \\ \vdots & \ddots & \vdots & \ddots & \vdots \\ Z_{N1} & \cdots & Z_{Nk} & \cdots & Z_{NN} \end{bmatrix} \begin{bmatrix} I_1 \\ \vdots \\ I_k \\ \vdots \\ I_N \end{bmatrix} \tag{3.46}$$

This means the kth equation can be eliminated from the above equation to leave with a reduced order impedance equation:

$$\begin{bmatrix} V_1 \\ \vdots \\ V_{k-1} \\ V_{k+1} \\ \vdots \\ V_N \end{bmatrix} = \begin{bmatrix} Z_{11} & \cdots & Z_{1k-1} & Z_{1k+1} & \cdots & Z_{1N} \\ \vdots & \ddots & \vdots & \vdots & \ddots & \vdots \\ Z_{k-11} & \cdots & Z_{k-1k-1} & Z_{k-1k+1} & \cdots & Z_{k-1N} \\ Z_{k+11} & \ddots & Z_{k+1k-1} & Z_{k+1k+1} & \ddots & Z_{k+1N} \\ \vdots & & \vdots & \vdots & & \vdots \\ Z_{N1} & \cdots & Z_{Nk-1} & Z_{Nk+1} & \cdots & Z_{NN} \end{bmatrix} \begin{bmatrix} I_1 \\ \vdots \\ I_{k-1} \\ I_{k+1} \\ \vdots \\ I_N \end{bmatrix} \tag{3.47}$$

One caution needs to be paid to the constant voltage bus is when the bus driving impedance is approaching zero (this can happen during transient stability simulation to a bus fault). Bus driving impedance is the corresponding diagonal element in Z_{bus} for that bus. Let us assume V_k is kept at a constant $\overline{V}_k$, and the corresponding driving point impedance z_{kk} is close to zero, under such a condition, by superposition theory, assuming all bus current injection zero except bus k, i.e. $I_i = 0 (i = 1, \ldots, N; i \neq k)$, following relationships from the impedance equation are obtained:

$$V_k = z_{kk} \times I_k, \quad \text{or} \quad I_k = \frac{\overline{V}_k}{z_{kk}} \to \infty \tag{3.48}$$

Therefore, a certain realistic constraint should be provided in adjacent to $V_k =$ constant condition, such as setting a range for I_k:

$$V_k = \overline{V}_k, \text{if } I_k \leq I_{\max} \tag{3.49}$$

3.3.5 Current Injection Components

Some devices can maintain nearly constant current characteristics on the output despite variations in voltage or load conditions. This is due to dedicated control functions or special electrical characteristics. These devices either supply currents into the network if they are sources or consume constant currents if they are loads. In general, they inject either positive or negative current into the connected buses and are referred to as current injection components. Examples of these components include power electronic circuits and components, some motor drives, LED drivers, battery chargers, arc welding power supplies, and electrochemical processes.

Buses with constant current injections are readily handled in the impedance equation by simply substituting the know current injection into the I vector on the right-hand side in Equation (3.38). In admittance equation, assume the voltage $\bar{I}_k$ at bus k is fixed, then the corresponding admittance equation becomes:

$$\begin{bmatrix} I_1 \\ \vdots \\ \bar{I}_k \\ \vdots \\ I_N \end{bmatrix} = \begin{bmatrix} Y_{11} & \cdots & Y_{1k} & \cdots & Y_{1N} \\ \vdots & \ddots & \vdots & \ddots & \vdots \\ Y_{k1} & \cdots & Y_{kk} & \cdots & Y_{kN} \\ \vdots & \ddots & \vdots & \ddots & \vdots \\ Y_{N1} & \cdots & Y_{Nk} & \cdots & Y_{NN} \end{bmatrix} \begin{bmatrix} V_1 \\ \vdots \\ V_k \\ \vdots \\ V_N \end{bmatrix} \tag{3.50}$$

Then, eliminate the kth equation to obtain a reduced order admittance equation in the form of:

$$\begin{bmatrix} I_1 \\ \vdots \\ I_{k-1} \\ I_{k+1} \\ \vdots \\ I_N \end{bmatrix} = \begin{bmatrix} Y_{11} & \cdots & Y_{1k-1} & Y_{1k+1} & \cdots & Y_{1N} \\ \vdots & \ddots & \vdots & \vdots & \ddots & \vdots \\ Y_{k-11} & \cdots & Y_{k-1k-1} & Y_{k-1k+1} & \cdots & Y_{k-1N} \\ Y_{k+11} & \ddots & Y_{k+1k-1} & Y_{k+1k+1} & \ddots & Y_{k+1N} \\ \vdots & \cdots & \vdots & \vdots & \cdots & \vdots \\ Y_{N1} & & Y_{Nk-1} & Y_{Nk+1} & & Y_{NN} \end{bmatrix} \begin{bmatrix} V_1 \\ \vdots \\ V_{k-1} \\ V_{k+1} \\ \vdots \\ V_N \end{bmatrix} \tag{3.51}$$

3.4 Model Integration

The control system models that govern the operation of the specific equipment, including machines, machine controls, converters, etc., shall be integrated together with equipment and network models to form a complete system model in order to fully cover system behaviors and dynamics and offer a simultaneous solution platform for different system studies and analysis.

In fact, assuming all series equipment and shunt equipment are linear and loads are constant impedance, those components can be directly built into the network Y-bus and Z-bus, as demonstrated in the previous sections. Constant voltage and constant current sources are also easy to handle in the admittance equation and the impedance equation, as discussed. Many loads are best described as constant power loads, meaning their power consumption roughly remains at constant levels when system operating conditions, e.g. voltage, are kept in a certain range. Some conversions need to be down in order to integrate these loads into the network questions. Let us consider bus k connecting a constant power load $S_k = P_k + jQ_k$, by realizing $(x)^*$ is the conjugate of complex variable x, bus injection current I_k can be written as:

$$I_k = \left(\frac{S_k}{V_k}\right)^* = \frac{P_k - jQ_k}{V_k^*} \tag{3.52}$$

Substitute it into the admittance equation to obtain,

$$\begin{bmatrix} I_1 \\ \vdots \\ \frac{P_k - jQ_k}{V_k^*} \\ \vdots \\ I_N \end{bmatrix} = \begin{bmatrix} Y_{11} & \cdots & Y_{1k} & \cdots & Y_{1N} \\ \vdots & \ddots & \vdots & \ddots & \vdots \\ Y_{k1} & \cdots & Y_{kk} & \cdots & Y_{kN} \\ \vdots & \ddots & \vdots & \ddots & \vdots \\ Y_{N1} & \cdots & Y_{Nk} & \cdots & Y_{NN} \end{bmatrix} \begin{bmatrix} V_1 \\ \vdots \\ V_k \\ \vdots \\ V_N \end{bmatrix} \tag{3.53}$$

Equation (3.53) is the modified admittance equation or can be called load flow equation which will be used in the load flow solution section. It is nonlinear due to the power term and thus cannot be solved directly. Iterative solution techniques will be discussed in the load flow solution section regarding how to solve it.

Equipment and devices that involve dynamic component or controls normally are expressed in combination of differential and algebra equations, or DAE (differential–algebraic equations). The general form of DAE can be written as:

$$\dot{x} = F(x, y, t) \tag{3.54}$$

$$0 = G(x, y, t) \tag{3.55}$$

where,

$\dot{x}$ is the derivative of the state variables of the component dynamic or controls.

$F(x, y)$ are the state space equations describing the dynamics equations or control functions.

y is the output variable of the state space equation.

$G(x, y)$ are the network equations, such as the admittance equation and load flow equation, and other algebra equations.

t is time.

Function $\dot{x} = F(x, y, t)$ could be simple SVC or converter control with the outputs are voltage or current or power injection to the network, or a combination of synchronous machine dynamic equation plus the machine prime mover/speed governor and excitation/AVR controls.

The control functions for the equipment and devices are integrated with the network equations in such a way that at each time instance, the state space equation is solved, and the state variables x and outputs y are updated. Both will be substituted into the network algebra equations to solve network variables such as bus voltage and branch currents.

Other dynamic equations for machines are also integrated with the network equations in the same manner.

3.5 Operation Modeling

I&C power systems have two major operation modes. One is in parallel with the utility, another is stand-alone or in an islanding model. Modeling these two modes appropriately is necessary for all types of analysis.

3.5.1 Utility Connected System

When there exists utility connection(s) to the local system, from the simulation point of view, one utility and its model should be selected as voltage regulator and reference machine (slack bus), meaning to assign its voltage magnitude and phase angle as fixed (typically the voltage magnitude equals the unity operating voltage and phase angle equals 0°) in load flow, motor starting, harmonic and transient stability analyses. For transient stability analysis, a reference rotor angle will be calculated and assigned to this utility based on its equivalent model. These processes are required for solving network and synchronous machine dynamic equations. The reference voltage angle and reference rotor angle will not change the solution in terms of voltage and current magnitude; they will only change the relative phase angles of these variables. As far as synchronous machines, each machines' relative rotor angle is always with respect to the reference angle, so they will not be changed, and the system stability assessment is not affected.

When a local system has multiple incoming line connections from either the same utility substation or from different substations, as shown in Figure 3.49, cautions should be paid to modeling each of the utility connecting point. In principle, any one of the utility connecting points can be chosen as the reference machine with a reference voltage phase angle, and reference rotor angle if it is dynamically modeled. The important modeling issue is about voltage angles of the rest utility connecting point voltages. Those voltage angles cannot be arbitrarily assigned; otherwise, erroneous results will be presented. Careful data collection procedures should be followed to obtain the voltage phase angle differences between all utility connecting points with respect to the reference utility and use them to assign the initial voltage angle for each utility connecting point. The true voltage phase angle for each utility connecting point is obtained using the reverse process.

Another note about the utility-connected system is that to realize all synchronous generators in the system are now in parallel operating with the utility.

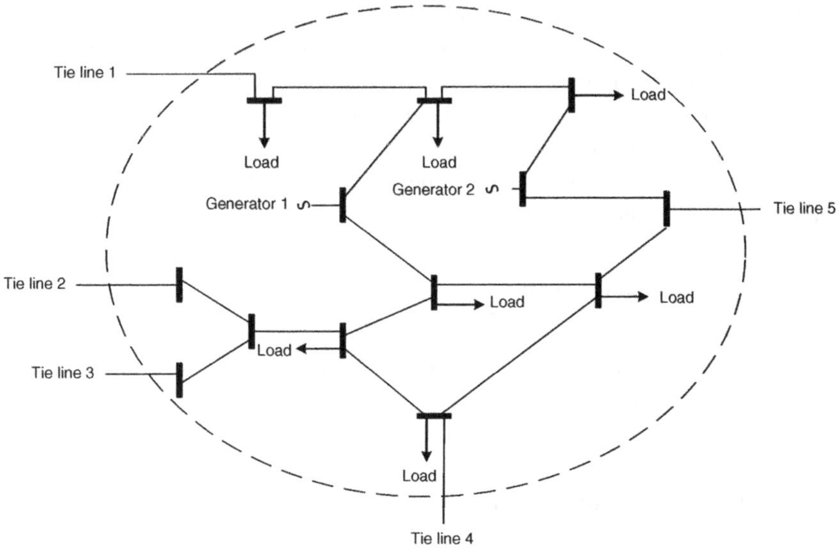

Figure 3.49 Multiple utility connection model.

So, in their governor control model settings, all speed control should be set as droop control, instead of isochronous control. This will make sure the simulation results match the real operating condition and truly reflect the machine speed in case of disturbances.

3.5.2 Island System

An island system refers to a local power system that has no utility connection. All power to the system loads is supplied by on-site generations. Still, at least one synchronous generator needs to be assigned as the reference machine for simulation purposes. This generator will have a choice to set its terminal bus voltage reference angle (so that the bus becomes the reference bus) for all types of studies except short-circuit study, which does not need a reference bus. The same generator will set as a reference rotor angle for transient stability studies. Usually, this generator is also set in isochronous speed control mode in transient stability study, so it is responsible for system frequency control. The other synchronous generators can either be set to isochronous speed control or, in most cases, set to droop control based on actual load-frequency control for each generator. In real operation, a group generators can be set to the same droop setting to allow share active power generation between them. Figure 3.50 illustrates an island system situation.

In transient stability study and analysis, system operation configuration can change from the utility connected to islanded as study scenarios. Once all initial

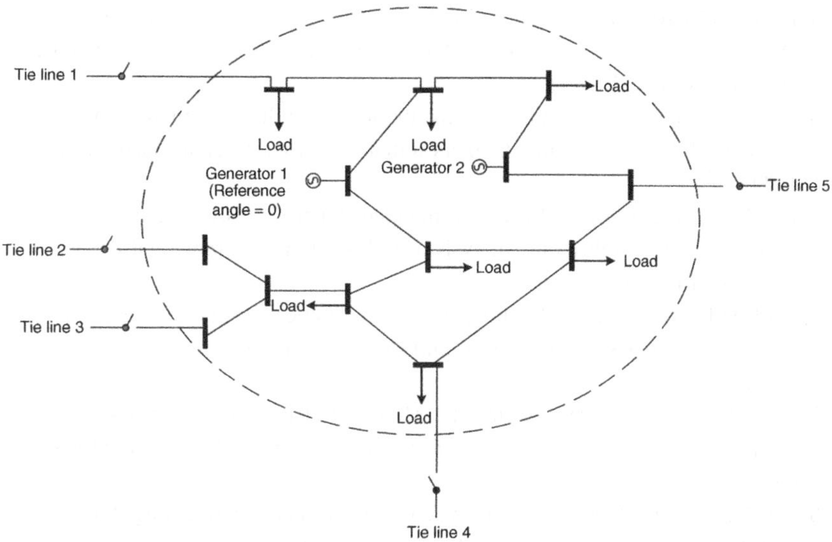

Figure 3.50 Island system reference bus and reference machine model.

voltage phase angles and rotor angle are set, there is no need to reset them when system configuration changes, as normally the computer model can automatically handle the transition. Nevertheless, when a portion of the system becomes islanded, it is still needed to have at least one generator to be set in isochronous speed control, otherwise the machine speed as well as the system frequency, will drift from the nominal value due to imbalance in generation and load demand. However, a isochronous generator will not always provide required active power to meet load demand under island situation. In that case, adding additional generation or reduce load demand, called load shedding, will be required.

References

1 IEEE Std 3002.3-2018, IEEE Recommended Practice for Conducting Short-Circuit Studies and Analysis of Industrial and Commercial Power Systems, IEEE, 2018.

2 IEEE C37.010-2016, IEEE Application Guide for AC High-Voltage Circuit Breakers > 1000 Vac Rated on a Symmetrical Current Basis, IEEE, 2016.

3 IEEE C37.13-1990, IEEE Standard for Low-Voltage AC Power Circuit Breakers Used in Enclosures, IEEE, 1990.

4 IEEE C37.013-1997, IEEE Standard for AC High-Voltage Generator Circuit Breakers Rated on a Symmetrical Current Basis, IEEE, 1997.

5 IEEE C37.4-2000, IEEE Standard Design Tests for High-Voltage Fuses, Distribution Enclosed Single-Pole Air Switches, Fuse Disconnecting Switches, and Accessories, IEEE, 2000.
6 IEEEC37.5-1979, IEEE Guide for Calculation of Fault Currents for Application of AC High-Voltage Circuit Breakers Rated on a Total Current Basis, IEEE, 1979.
7 IEEE Std 3002.7-2018, IEEE Recommended Practice for Conducting Motor-Starting Studies and Analysis of Industrial and Commercial Power Systems, IEEE, 2018.
8 IEEE Std 3002.8-2018, IEEE Recommended Practice for Conducting Harmonic Studies and Analysis of Industrial and Commercial Power Systems, IEEE, 2018.
9 Dai, J.J. (2012). Chapter 9 Computer Modeling of Rotating Machines. In: *Electromagnetic Transients in Transformer and Rotating Machine Windings* (ed. C.Q. Su). IGI Global.
10 IEEE1110-2019, IEEE Guide for Synchronous Generator Modeling Practices and Parameter Verification with Applications in Power System Stability Analyses, IEEE, 2019.
11 Kundur, P. (1994). *Power System Stability and Control*. McGraw-Hill, Inc.
12 IEEE 399-1997, IEEE Recommended Practice for Industrial and Commercial Power Systems Analysis, IEEE, 1997.
13 Singh, B., Bhuvaneswari, G., and Garg, V. (2006). Harmonic Mitigation Using 12-Pulse AC–DC Converter in Vector-Controlled Induction Motor Drives. *IEEE Transactions On Power Delivery* 21 (3): 1483–1492.
14 J.J Dai et al. “A user’s Guide to Selecting and Applying Medium Voltage Adjustable Speed Drives”, 2010 Record of Conference Papers Industry Applications Society 57th Annual Petroleum and Chemical Industry Conference (PCIC), 20–22 September. 2010.
15 IEEE Std 141-1999, Recommended Practice for Electric Power Distribution for Industrial Plants, IEEE, 1999.
16 IEEE, Dynamic Models for Turbine-Governors in Power System Studies, TECHNICAL REPORT PES-TR1, Jan. 2013.
17 IEEE Std 421.5-2016, IEEE Recommended Practice for Excitation System Models for Power System Stability Studies, IEEE, 2016.
18 IEEE 100-2000, The Authoritative Dictionary of IEEE Standards Terms Seventh Edition, IEEE, 2000.
19 IEEE Special Stability Controls Working Group, “Static VAr compensator models for power flow and dynamic performance simulation”, *IEEE Transactions on Power Systems*, Volume: 9, Issue: 1, Feb. 1994, Page(s): 229–240.

4

Industrial and Commercial Power System Mathematical Solutions

Industrial and commercial (I&C) power system analysis relies on mathematical solutions to system models which integrate the network components and source and load components together with controls using various numerical computational algorithms. Each type of analysis requires specific and most efficient solutions, since the models are not exactly the same for different analyses. For example, load flow analysis basically solves the admittance equation since all network, source and load components are well integrated into the Y-bus. Load flow solution is about the steady-state of the network, meaning at a given operating point with sources and loads are fixed, except that generator automatic voltage regulator (AVR), transformer tap-changer, capacitor, and SVC compensator setting points are adjustable based on the specified control configuration and control objectives. There is no time variable involved in the steady state calculation. Short-circuit solution, on the other hand, is more suitable for using the impedance equation to formulate the problem, as the fault current can be readily calculated if the Z-bus is known. Short-circuit solution does not entirely deal with a steady state, since there are several fault duty currents needed to be considered for breaker capabilities assessment. Therefore, short-circuit solution will need to construct a number of Z-bus to represent different instants of time for analysis interests. Likewise, motor starting, harmonic and transient stability solutions need either to modify system base admittance and impedance equations or append additional conditions to the complete system equations to describe the study dependent conditions and constraints to obtain complete, accurate, and efficient solutions.

4.1 Load Flow Solution

The load flow solution process is based on the system admittance equation with sources and loads specified to solve nodal, i.e. bus voltages. Power system

Industrial and Commercial Power System Analysis Fundamentals and Practice, First Edition. J. J. Dai.

bus voltages are state variables, meaning once all bus voltages are determined, branch currents and power flows, source and loads currents and powers then can be uniquely derived, and the entire power system solution is obtained. There are quite a few load flow solution algorithms available and most of them are implemented in various computer software packages. Based on the approximation approach and convergence rate, the load flow solution algorithms are classified as direct iteration solution, the first-order solution and the higher-order solution.

4.1.1 Direct Iteration Solution

Recall the admittance equation discussed in Chapter 3,

$$\begin{bmatrix} I_1 \\ \vdots \\ I_k \\ \vdots \\ I_N \end{bmatrix} = \begin{bmatrix} Y_{11} \cdots Y_{1k} \cdots Y_{1N} \\ \vdots \quad \ddots \quad \vdots \quad \ddots \quad \vdots \\ Y_{k1} \cdots Y_{kk} \cdots Y_{kN} \\ \vdots \quad \ddots \quad \vdots \quad \ddots \quad \vdots \\ Y_{N1} \cdots Y_{Nk} \cdots y_{NN} \end{bmatrix} \begin{bmatrix} V_1 \\ \vdots \\ V_k \\ \vdots \\ V_N \end{bmatrix} \tag{4.1}$$

Without losing generality, assume the constant voltage components and current inject components are already handled through substitutions into and eliminations from the original admittance equation, and only constant power components are to be considered. Let the load power $S_k = P_k + jQ_k$ at bus k be specified, the bus k injection current then can be formulated as,

$$I_k = \left(\frac{\overline{S}_k}{V_k} \right)^* = \frac{\overline{P}_k - j\overline{Q}_k}{V_k^*} = Y_{k1}V_1 + \cdots + Y_{kk}V_k + \cdots + Y_{kN}V_N \tag{4.2}$$

This formulation will make the admittance equation expressed in terms of connected powers at nodes and buses, thus it is referred to as power flow equation. The power flow equation is a nonlinear algebra equation which is typically solved using iteratively by numerical methods. The direct iteration solution of power flow equation refers to an approximation process without involving variable derivatives. Here are the basic steps involved in obtaining a direct iterative solution in load flow analysis. This solution technique is also well known as Gaussian iteration method.

First rearrange Equations (4.1) and (4.2) and generalize it for every bus in the network,

$$V_k = \frac{1}{Y_{kk}} \left[\frac{\overline{P}_k - j\overline{Q}_k}{V_k^*} - (Y_{k1}V_1 + \cdots + Y_{kk-1}V_{k-1} + Y_{kk+1}V_{k+1} + \cdots + Y_{kN}V_N) \right]$$

$$(k = 1, \ldots, N) \tag{4.3}$$

The above equations can be solved iteratively following the steps below:

1) Decide a set of initial values for bus voltages, typically the nominal voltage in 1.0 per unit with 0 phase angle.
2) Set bus voltages to initial values and construct a bus voltage vector $V_{bus}^{(0)}$, where 0 stands for the iteration step 0 or the initial values.
3) Starting from iteration number $l = 1$, let all voltages $V_k^{(l-1)}$ on the right-hand side of Equation (4.3) be the previous value and solve for the next step $V_k^{(l)}$ for all buses (except those are with constant and fixed values) sequentially and iteratively. The calculation is carried out for all unknown bus voltages until either the maximum iteration number L_{max} is reached or a specified convergence criteria are met,

$$V_k^{(l)} = \frac{1}{Y_{kk}} \left[\frac{\overline{P}_k - j\overline{Q}_k}{V_k^{(l-1)*}} - \left(Y_{k1} V_1^{(l-1)} + \cdots + Y_{kk-1} V_{k-1}^{(l-1)} + Y_{kk+1} V_{k+1}^{(l-1)} + \cdots + Y_{kN} V_N^{(l-1)} \right) \right] \ (k = 1, \ldots, N), \\ (l = 1, \ldots, L_{max} \text{ or calculation converges}) \tag{4.4a}$$

Or to improve the convergence, assuming the iteration follows the order $k = 1, \ldots, N$ in solving $V_k^{(l)}$, then the newly solved $V_i^{(l)}$ $(i = 1, \ldots, k-1)$ can be substituted into Equation (4.4a) to calculate $V_k^{(l)}$:

$$V_k^{(l)} = \frac{1}{Y_{kk}} \left[\frac{\overline{P}_k - j\overline{Q}_k}{V_k^{(l-1)*}} - \left(Y_{k1} V_1^{(l)} + \cdots + Y_{kk-1} V_{k-1}^{(l)} + Y_{kk+1} V_{k+1}^{(l-1)} + \cdots + Y_{kN} V_N^{(l-1)} \right) \right] \ (k = 1, \ldots, N), \\ (l = 1, \ldots, L_{max} \text{ or calcualtion converges}) \tag{4.4b}$$

The convergence criteria are usually set to check bus voltage deviation:

$$\left| \left| V_k^{(l)} \right| - \left| V_k^{(l-1)} \right| \right| \le \varepsilon_V \text{ for all } k = 1, \ldots, N \tag{4.5}$$

or to check bus power injection deviations for those buses with P and Q specified:

$$\left| P_k^{(l)} - \overline{P}_k \right| \le \varepsilon_P \text{ for all } k = 1, \ldots, N \text{ with } P \text{ specified} \tag{4.6a}$$

$$\left| Q_k^{(l)} - \overline{Q}_k \right| \le \varepsilon_Q \text{ for all } k = 1, \ldots, N \text{ with } Q \text{ specified} \tag{4.6b}$$

where $\left| V_k^{(l)} \right|$ is the magnitude of phasor $V_k^{(l)}$, $P_k^{(l)}$, and $Q_k^{(l)}$ are computed from the bus voltage and injection current, and ε_V, ε_P, and ε_Q are predetermined thresholds for voltage, active power, and reactive power, respectively.

4.1.2 First-Order Solution

By linearizing the power flow equation around the initial conditions of the state variables, i.e. bus voltage magnitudes and angles, a set of linear equations can be obtained. The linearization involves finding the first-order derivatives of the power flow equation with respect to the state variables. The linearized equations are then solved for the new state variables, which will be applied to perform a new linearization. The above process will repeat until the state variables converge to final stable values. This load flow solution method is referred to as Newton–Raphson method.

Start from the following power balance equation at bus k developed previously:

$$P_k - jQ_k = V_k^*(Y_{k1}V_1 + \cdots + Y_{kk}V_k + \cdots + Y_{kN}V_N) \tag{4.7}$$

or

$$\begin{aligned} P_k &= Re\left[V_k^*(Y_{k1}V_1 + \cdots + Y_{kk}V_k + \cdots + Y_{kN}V_N)\right] \\ &= p_k(|V_1|, \ldots, |V_N|, \delta_1, \ldots, \delta_N) \end{aligned} \tag{4.8a}$$

$$\begin{aligned} Q_k &= Im\left[-V_k^*(Y_{k1}V_1 + \cdots + Y_{kk}V_k + \cdots + Y_{kN}V_N)\right] \\ &= q_k(|V_1|, \ldots, |V_N|, \delta_1, \ldots, \delta_N) \end{aligned} \tag{4.8b}$$

where $p_k(\cdot)$ and $q_k(\cdot)$ represent of a function of $(|V_1|, \ldots, |V_N|, \delta_1, \ldots, \delta_N)$, $|V_k|$ is the bus voltage magnitude, δ_k is the bus voltage angle noting in general a bus voltage can be expressed as,

$$V = |V|e^{j\delta} \tag{4.9}$$

Linearize Equations (4.8a) and (4.8b) to obtain small variations of bus powers by leveraging the first-order approximations of functions $p_k(\cdot)$ and $q_k(\cdot)$,

$$\Delta P_k = \frac{\partial p_k}{\partial |V_1|}\Delta |V_1| + \ldots + \frac{\partial p_k}{\partial |V_N|}\Delta |V_N| + \frac{\partial p_k}{\partial |\delta_1|}\Delta \delta_1 + \ldots + \frac{\partial p_k}{\partial |\delta_1|}\Delta \delta_N \tag{4.10a}$$

$$\Delta Q_k = \frac{\partial q_k}{\partial |V_1|}\Delta |V_1| + \ldots + \frac{\partial q_k}{\partial |V_1|}\Delta |V_N| + \frac{\partial q_k}{\partial |\delta_1|}\Delta \delta_1 + \ldots + \frac{\partial q_k}{\partial |\delta_1|}\Delta \delta_N \tag{4.10b}$$

or write them in a matrix form,

$$\begin{bmatrix} \Delta P \\ \Delta Q \end{bmatrix} = \begin{bmatrix} \dfrac{\partial p_k}{\partial |V|} & \dfrac{\partial p_k}{\partial |\delta|} \\ \dfrac{\partial q_k}{\partial |V|} & \dfrac{\partial q_k}{\partial |\delta|} \end{bmatrix} \begin{bmatrix} \Delta |V| \\ \Delta \delta \end{bmatrix} = [J] \begin{bmatrix} \Delta |V| \\ \Delta \delta \end{bmatrix} \tag{4.11}$$

where ΔP, ΔQ, $\Delta |V|$, and $\Delta\delta$ all have the network dimensions, $[J]$ is called Jacobian matrix and should be inversible unless the system equation is in an ill-condition.

Starting from l = 0 to the maximum iteration number L_{max}, Equation (4.11) is solved iteratively for bus voltages by taking the inverse of [***J***],

$$\begin{bmatrix} \Delta|V^{(l)}| \\ \Delta\delta^{(l)} \end{bmatrix} = [J^{(l)}]^{-1} \begin{bmatrix} \Delta P^{(l)} \\ \Delta Q^{(l)} \end{bmatrix} \tag{4.12}$$

To start, bus voltages are initialized to their nominal values with zero-phase angle unless there are better guesses. At each iteration step, bus power or load mismatches are calculated and new bus voltages are updated which are to be used to recalculate the Jacobian matrix,

$$\Delta P^{(l+1)} = \overline{P} - P^{(l)} \tag{4.13a}$$

$$\Delta Q^{(l+1)} = \overline{Q} - Q^{(l)} \tag{4.13b}$$

$$|V^{(l+1)}| = |V^{(l)}| + \Delta|V^{(l)}| \tag{4.14a}$$

$$\delta^{(l+1)} = \delta^{(l)} + \Delta\delta^{(l)} \tag{4.14b}$$

$$[J^{(l+1)}] = \begin{bmatrix} \dfrac{\partial p_k}{\partial |V|} & \dfrac{\partial p_k}{\partial |\delta|} \\ \dfrac{\partial q_k}{\partial |V|} & \dfrac{\partial q_k}{\partial |\delta|} \end{bmatrix}_{(|V|=|V^{(l+1)}|,\delta=\delta^{(l+1)})} \tag{4.15}$$

The Newton–Raphson load flow solution usually checks the following bus power or load mismatches for terminating the convergence:

$$\left|P_k^{(l)} - \overline{P}_k^{(l-1)}\right| \le \varepsilon_P \text{ for all } k = 1, \dots, N \tag{4.16a}$$

$$\left|Q_k^{(l)} - \overline{Q}_k^{(l-1)}\right| \le \varepsilon_Q \text{ for all } k = 1, \dots, N \tag{4.16b}$$

The first-order load flow solutions generally provide faster convergence as well as higher accuracy. It should be mentioned that since they need to involve derivatives of functions, bad initial guesses and ill load flow equation conditions can cause divergence in the numerical computation. Enhanced convergence and improved computation techniques have been developed and implemented in some computer software packages.

4.1.3 Higher-Order Solution

Besides the direct iteration and the first-order solutions, there are also the second-order solutions such as those based on current injection equations [1]. These methods utilize the second-order derivatives of the load flow equations, called Hessian matrix [***H***] as shown in Equation (4.17). The second-order load flow methods have promising advantages to offer a faster solution in terms of reducing power mismatches.

$$\begin{bmatrix} \Delta P \\ \Delta Q \end{bmatrix} = [J] \begin{bmatrix} \Delta |V| \\ \Delta \delta \end{bmatrix} + \frac{1}{2}[H] \begin{bmatrix} [\Delta |V|]^T [\Delta |V|] \\ [\Delta \delta]^T [\Delta \delta] \end{bmatrix} \tag{4.17}$$

4.2 Short-Circuit Solution

Short-circuit analysis is to evaluate the system's response to fault conditions by calculating fault currents at specified network locations. The calculated fault duty currents are used for setting and adjustment of relays, circuit breakers, fuses, and other protection devices in system design and validation. Normally, these locations are the busbars, equipment terminals, joint connections of components, or anywhere energized conductors and equipment are exposed and faults between these parts or parts to ground can occur. Short-circuit faults being considered in I&C power systems are three-phase fault, single-line-to-ground (L–G) fault, line-to-line (L–L) fault, or double-line-to-ground (L–L–G) fault. These are all parallel faults, or shunt faults, to be distinguished from series faults which refer to open line faults that often take place in transmission and distribution systems. Except for the three-phase fault, all other faults are known as unbalanced faults.

Commonly calculated fault types consist of three-phase bold fault, L–L fault, L–L–G fault, and L–G fault. Configurations of these faults are illustrated in Figure 4.1.

4.2.1 Balanced Fault Solution

A balanced fault in I&C power systems refers to a fault condition where the three phase currents and voltages are in same magnitude and 120° out of phase from each other. Dealing with balanced faults, Z_{bus} can be applied directly.

Assuming bus k is faulted in three-phase, from the bus impedance equation,

$$\begin{bmatrix} \overline{E}_1 \\ \vdots \\ \overline{E}_k \\ \vdots \\ \overline{E}_N \end{bmatrix} = \begin{bmatrix} Z_{11} & \cdots & Z_{1k} & \cdots & Z_{1N} \\ \vdots & \ddots & \vdots & \ddots & \vdots \\ Z_{k1} & \cdots & Z_{kk} & \cdots & Z_{kN} \\ \vdots & \ddots & \vdots & \ddots & \vdots \\ Z_{N1} & \cdots & Z_{Nk} & \cdots & Z_{NN} \end{bmatrix} \begin{bmatrix} 0 \\ \vdots \\ I_k \\ \vdots \\ 0 \end{bmatrix} \tag{4.18}$$

where all nonfaulted buses have 0 bus injection current except bus k has the fault current I_k. All fault current sources including utilities, generators, and motors are presented by their Thevenin equivalent as a constant voltage source behind an impedance as shown in Figure 4.2. All impedance of the Thevenin equivalent is already included in the Z_{bus}.

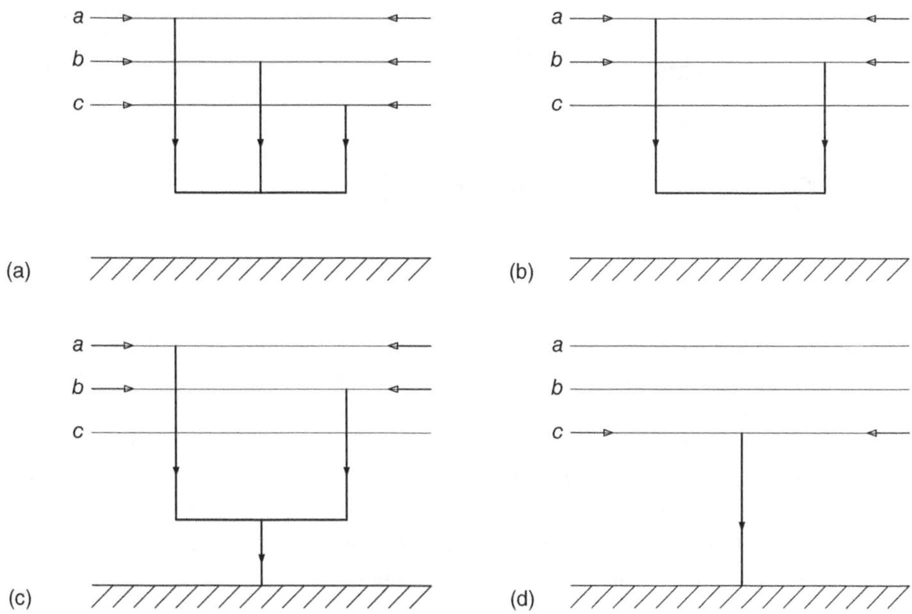

Figure 4.1 Examples of different short-circuit faults. (a) Balanced three-phase fault, (b) two-phase ungrounded fault, (c) two-phase to grounded fault, and (d) single-line to ground fault.

Figure 4.2 Thevenin equivalent of fault current source.

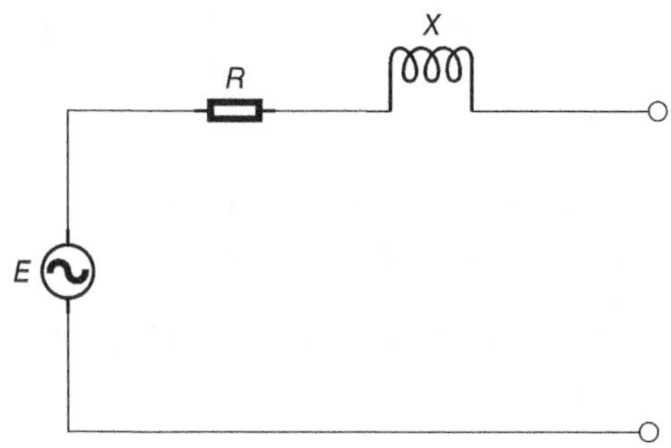

Inverse Equations (4.18) to solve the fault current I_k,

$$I_{fault} = I_k = \text{the } kth \text{ row of } \left([Z_{bus}]^{-1} \begin{bmatrix} \overline{E}_1 \\ \vdots \\ \overline{E}_k \\ \vdots \\ \overline{E}_N \end{bmatrix} \right) \tag{4.19}$$

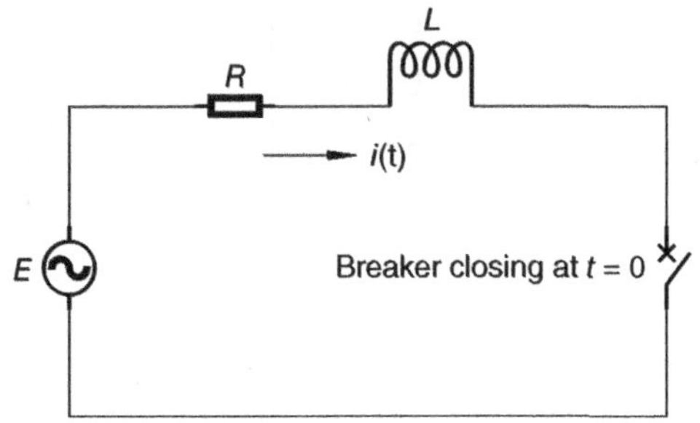

Figure 4.3 Circuit model with steady-state ac current sources. Source: IEEE 3002.3–2018 [2]/with permission of IEEE.

If it is a three-phase impedance fault, the fault impedance should be built into the Z_{bus}, and the solution formulation remains the same.

Due to the transient nature and source current decay, I_{fault} is not a constant value. This can be seen by examining the following equivalent circuit diagram Figure 4.3 consisting of an ac source, an equivalent resistance R, and an inductance L. The total fault current $i(t)$ after the switch is closed can be expressed as a summation of three components: i.e. decaying dc current, decaying ac current, and steady-state ac current as shown in Equation (4.20).

$$i(t) = i_{dc\ decay}(t) + i_{ac\ decay}(t) + i_{ac\ steady\ state}(t) \tag{4.20}$$

Alternatively, $i(t)$ can also be expressed as machine model parameters in Equation (4.21),

$$i(t) = E\left[\frac{1}{X_d} + \left(\frac{1}{X'_d} - \frac{1}{X_d}\right)e^{-\frac{t}{T'_d}} + \left(\frac{1}{X''_d} - \frac{1}{X'_d}\right)e^{-\frac{t}{T''_d}}\right] \tag{4.21}$$

where,

E is the rms value of the machine open circuit voltage.
X_d, X'_d, X''_d are the direct-axis synchronous, transient, and subtransient reactances, respectively.
T'_d, T''_d are the direct-axis transient and subtransient short-circuit time constants, respectively.

Generally, the fault currents in I&C power systems are asymmetrical as shown in Figure 4.4.

Due to the nature that fault current decaying with time, fault currents at three stages are selected by industry standards for circuit breaker and fuse rating and relay setting checkup. These are,

1) I_{mom} is the fault current within one cycle immediately after the occurrence of the fault. I_{mom} represents "first-cycle" duty, or momentary duty which is compared to circuit breaker closing and latching capabilities.

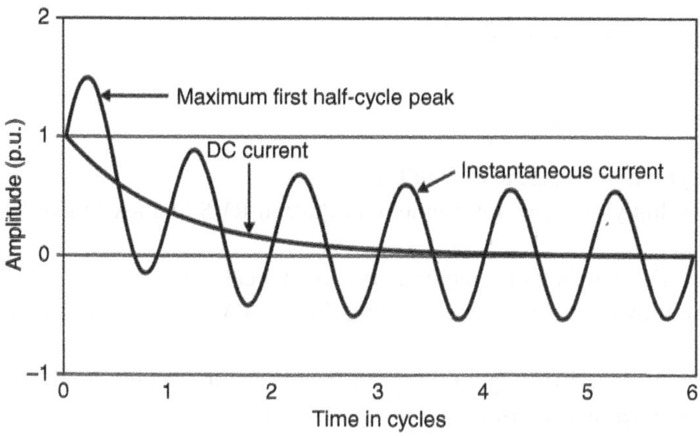

Figure 4.4 Asymmetrical ac short-circuit current. Source: IEEE 3002.3-2018 [2]/with permission of IEEE.

2) I_{int} is the fault current during the period from 1.5 to 4 cycles after the fault. I_{int} represents interrupting duty relevant to medium- and high-voltage circuit breaker parting times, which is compared to circuit breaker interrupting or breaking capabilities.
3) $I_{steady\ state}$ is the steady state short-circuit current beyond 30 cycles from the moment of the fault inception. $I_{steady\ state}$ is used for back-up time-delay relays.

General formula for these currents in RMS, peak, symmetrical, asymmetrical (or adjusted) values are derived from the corresponding impedance matrix Z_{bus} constructed according to Tables 3.5 and 3.9 as well as other important parameters and are listed below,

$$I_{mom,rms,symm} = \frac{V_{pre-fault}}{\sqrt{3}Z^{eq}_{bus,mom}} \tag{4.22}$$

$$I_{mom,rms,asymm} = MF_{mom} \times I_{mom,rms,symm} \tag{4.23}$$

$$I_{mom,peak} = MF_{peak} \times I_{mom,rms,symm} \tag{4.24}$$

$$I_{int,rms,symm} = \frac{V_{pre-fault}}{\sqrt{3}Z^{eq}_{bus,int}} \tag{4.25}$$

$$I_{int,rms,adj} = MF_{adj} \times I_{int,rms,symm} \tag{4.26}$$

$$I_{steady\ state,rms} = \frac{V_{pre-fault}}{\sqrt{3}Z^{eq}_{bus,stead\ state}} \tag{4.27}$$

where,

$V_{pre-fault}$ is the pre-fault voltage at the fault location.
MF_{mom} is the multiplying factor for momentary duty in RMS and asymmetrical value.
MF_{peak} is the multiplying factor for momentary duty in peak value.
MF_{adj} is the multiplying factor for interrupting duty in RMS and asymmetrical value.
$Z^{eq}_{bus,mom}$ is the equivalent driving point impedance at the fault location from Z_{bus} based on the momentary (first-cycle) impedance network.
$Z^{eq}_{bus,int}$ is the equivalent driving point impedance at the fault location from Z_{bus} based on the interrupting impedance network.
$Z^{eq}_{bus,stead\ state}$ is the equivalent driving point impedance at the fault location from Z_{bus} based on the steady state impedance network.

All multiplying factors depend on X/R ratio and some other factors of the corresponding Z^{eq}_{bus}.

Thus, using Figure 4.3 equivalent circuit for each source, let the equivalent (internal) voltage stays at constant and its impedance varies corresponding to three fault stages, all fault duties can be calculated respectively. Many detailed tables, equations, multiplying and adjustment factors regarding the calculations are found in the relevant industrial standards [2–6], e.g. IEEE 3002.3, IEEE 399, IEEE Violet Book, and IEEE C37 series. This book focuses on the general solution and analysis approaches.

4.2.2 Unbalanced Fault Solution

For the unbalanced faults, three phase voltages and currents lose symmetry, which requires a special technique of symmetrical component transformation to handle the fault current calculation.

Using X to denote V or I variable, the symmetrical component method can transform forward and backward the voltage and current variables between the original balanced three-phase frame (X_a, X_b, X_c) and the sequence frame (X_0, X_1, X_2) through the following operations,

$$\begin{bmatrix} X_0 \\ X_1 \\ X_2 \end{bmatrix} = \frac{1}{3} \begin{bmatrix} 1 & 1 & 1 \\ 1 & a & a^2 \\ 1 & a^2 & a \end{bmatrix} \begin{bmatrix} X_a \\ X_b \\ X_c \end{bmatrix} \tag{4.28}$$

$$\begin{bmatrix} X_a \\ X_b \\ X_c \end{bmatrix} = \begin{bmatrix} 1 & 1 & 1 \\ 1 & a^2 & a \\ 1 & a & a^2 \end{bmatrix} \begin{bmatrix} X_0 \\ X_1 \\ X_2 \end{bmatrix} \tag{4.29}$$

where $a = e^{j\frac{2\pi}{3}}$ denotes a phase-shift operation of +120° (clockwise).

Frame (X_0, X_1, X_2) is defined such that,

1) X_0 is the zero sequence variable. It is of equal magnitude in all three phases.
2) X_1 is the system of balanced positive sequence, of equal magnitude in each phase, and of the same phase sequence as the original system, i.e. phase b lagging phase a by 120°, and phase c leading phase a by 120°.
3) X_2 is the system of balanced negative sequence, of equal magnitude in each phase, and of the opposite phase sequence as the original system, i.e. phase c lagging phase a by 120°, and b leading a by 120°.

To apply this method in the unbalanced short-circuit fault current calculation, three sets of networks, namely, positive sequence, negative sequence, and zero sequence networks, need to be configured and used to construct corresponding Z_{bus} matrices. Sequence networks need to take into consideration of component impedances in positive, negative, and zero sequences, as well as the component winding connections (whether in Y or Δ configuration), grounding (grounded or ungrounded), and grounding impedances. Figure 4.5 illustrates an example of how the sequence networks look for a three-phase system by incorporating the above conditions.

Using the sequence networks, fault currents for different unbalanced faults can be found as:

Single-line-to-ground fault (L–G). The faulted phase current is,

$$I_{fault} = \frac{3V_{pre-fault}}{Z_1 + Z_2 + Z_0 + 3Z_f} \tag{4.30}$$

Line-to-line fault (L–L). One of the faulted phase currents is,

$$I_{fault} = \frac{-j\sqrt{3}V_{pre-fault}}{Z_1 + Z_2 + Z_f} \tag{4.31}$$

Another faulted phase current is in the opposite phase direction.

Double-line-to-ground fault (L–L–G). The fault current to the ground is,

$$I_{fault} = \frac{V_{pre-fault}}{Z_1 + \frac{Z_2(Z_0+3Z_f)}{Z_2+Z_0+Z_f}} \tag{4.32}$$

In the above equations, Z_f is the fault impedance (set Z_f to 0 if it is a bold fault).

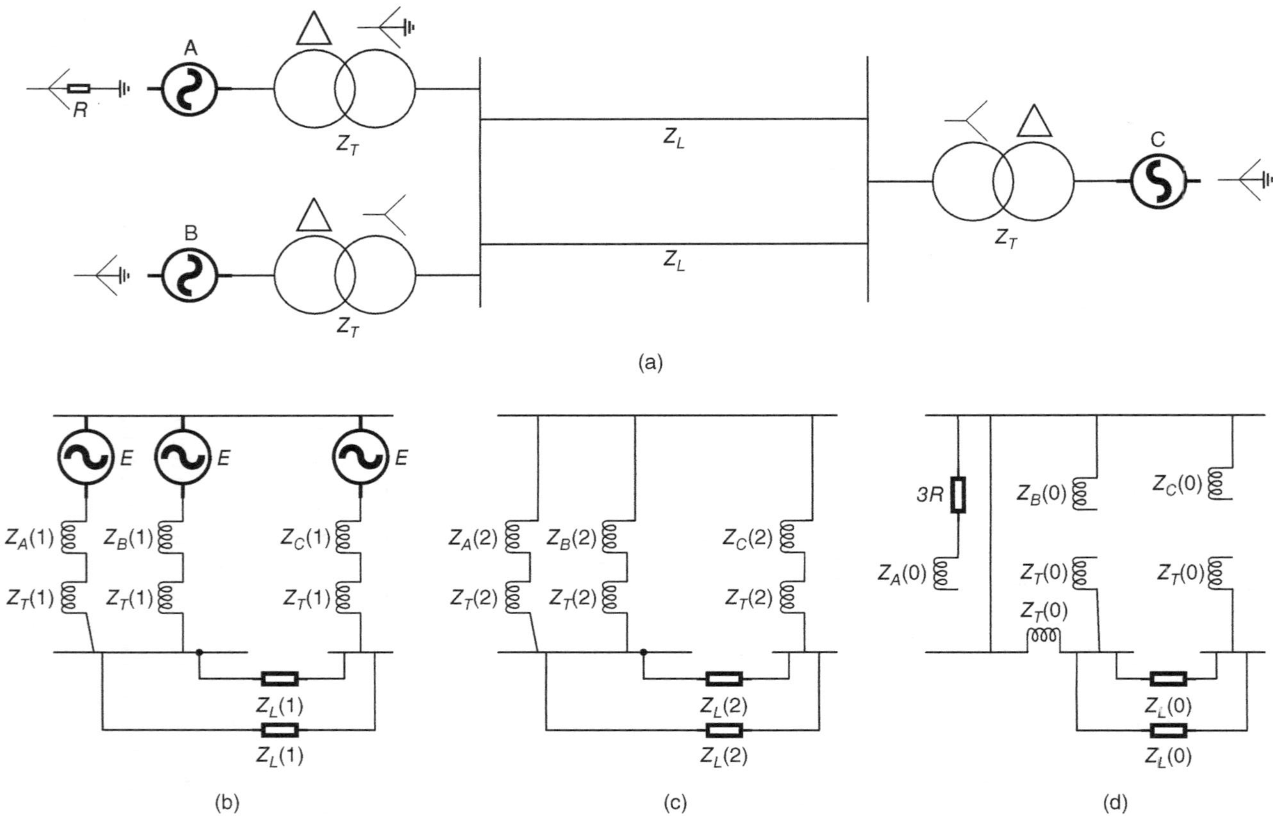

Figure 4.5 Example of sequential networks. (a) System single line diagram, (b) positive sequence net, (c) negative sequence net, and (d) zero sequence net.

4.3 Motor Starting Solution

Motor starting solution is derived from system equations (matrices), source models and starting motor models. Depending on the level of detail in both the motor models and the source models, motor starting solutions can be further categorized into static model solution and dynamic model solution. In all solution approaches, the network loads for the entire system will be included in the simulation, and utility and machine (including generators and motors) impedances and internal voltages are also required. Therefore, the motor starting solutions require both load flow solution and short-circuit solution data to perform the calculations.

4.3.1 Static Solution

The static motor starting solution refers to the modeling assumption that the starting motor is not dynamically modeled, i.e. no time variant variables nor differential equations are involved in the simulation. Rather, the starting motor is simplified as a constant impedance load during the acceleration. It is then eventually converted to a constant power load afterwards for the rest of the acceleration time. So, the static motor starting solution is divided into two stages:

Constant Impedance Stage. This stage corresponds to the initial moment of the motor starting in which the motor is dominated by its locked rotor current or in-rush current I_{LRC} (normally five to seven times of the motor rated current). An equivalent locked rotor impedance Z_{LR} in Equation (4.33) for the motor is calculated as,

$$Z_{LR} = \frac{V_{L-L}}{\sqrt{3}I_{LRC}}\ (\Omega) \tag{4.33}$$

Z_{LR} is to replace the starting motor as shown in Figure 4.6 and is included into the network admittance equation to solve the system voltages at all buses by

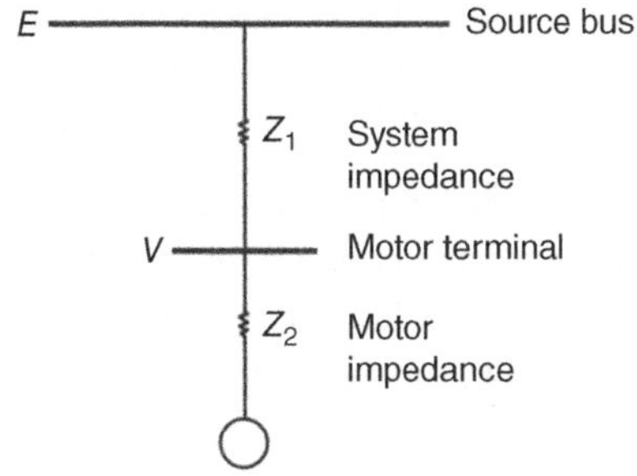

Figure 4.6 Simplified starting motor constant impedance model. Source: IEEE Std 3002.7™-2018 [7]/with permission of IEEE.

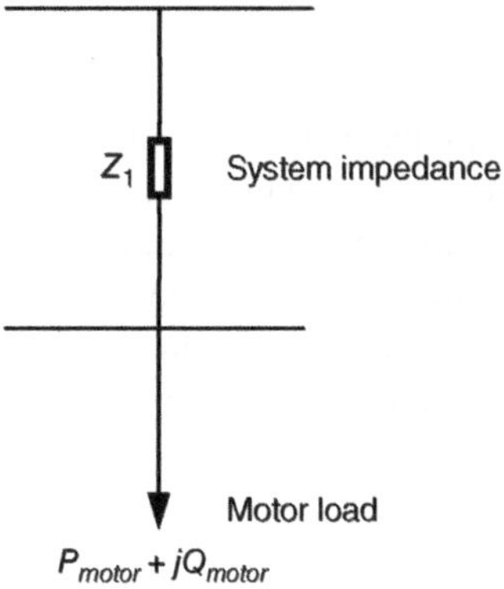

Figure 4.7 Simplified starting motor constant power model.

load flow solution. The solution represents the initial system voltages and voltage drops due to the small impedance of the motor at the initial locked rotor condition.

Constant Power Stage. In reality, once a motor is started and reaches its rated speed, the in-rush current will sharply drop close to the rated operating current depending on the loading condition and the motor will exhibit as a constant power load. Then, the motor model needs to convert from constant impedance to constant power as shown in Figure 4.7. The motor powers are calculated according to its loading from Equations (4.34) and (4.35).

$$P_{motor} = \text{Loading}(\%) \times PF_{rated} \times S_{rated} \tag{4.34}$$

$$Q_{motor} = \text{Loading}(\%) \times \sqrt{1 - PF_{rated}^2} \times S_{rated} \tag{4.35}$$

where,

P_{motor} is the active power.

P_{motor} is the reactive power.

PF_{rated} is the motor rated power factor.

S_{rated} is the motor rated apparent power.

Loading(%) is the motor loading percentage.

The switching time from the constant impedance stage to the constant power stage is based on an estimated motor acceleration time t_a derived from Equation (4.36). The load flow solution then is carried out to handle the motor constant power load as described in Chapter 3.

$$t_a = \frac{WR^2(s_{final} - s_{initial})}{308T_{net}} \text{ (sec.)} \tag{4.36}$$

where,

WR^2 is the motor moment of inertia in lb/ft^2 or kg/m^2.
s_{final} is the motor final speed in rev/min (RPM).
$s_{initial}$ is the motor initial speed in rev/min (RPM).
T_{net} is the average net acceleration torque equal to motor torque minus load torque N/m.

The load flow solution is applied to the stage 2 model and solves system voltages for the conditions until the motor acceleration is completed. Then the running motor is added to the system.

Static motor starting solution captures the worst voltage dip condition at the moment of motor starting, and the final system condition after the motor is fully started. It does not provide any system and starting motor details during the starting period including motor and load torque, acceleration torque, motor starting current, and motor speed. The acceleration time is also estimated which could be over or less than the practical situation. To obtain a more detailed and accurate solution, dynamic motor starting solution needs to be engaged.

4.3.2 Dynamic Solution

Dynamic motor starting solution uses motor equivalent circuit model (Figure 3.9) and load torque-slip curve (Figure 3.19) together to enable both motor and load dynamics into consideration for simulation and provide more realistic motor starting solution profile.

Starting from a reference time determined by the motor starting moment, a progressive data assembling and calculation update begins, which will continue to move to the next time step once the system voltages are solved at the current time. At each time step, the motor torque T_m is computed based on the present stator voltage at the motor terminal from the specified motor equivalent circuit,

$$T_m = \frac{3V^2\frac{R_r}{s}}{\omega_s\left[\left(R_s + \frac{R_r}{s}\right)^2 + (X_s + X_r)^2\right]} \tag{4.37}$$

where,

V_s is the motor stator voltage.
R_s is the motor stator resistance.
R_r is the motor rotor resistance.
X_s is the motor stator reactance.
X_m is the motor air gap (magnitizing) reactance.
s is the motor slip.
ω_s is the synchronous angular speed.

From the above motor torque T_m and motor load torque T_l from the load torque-slip curve Figure 4.8 at the current motor speed, motor acceleration torque

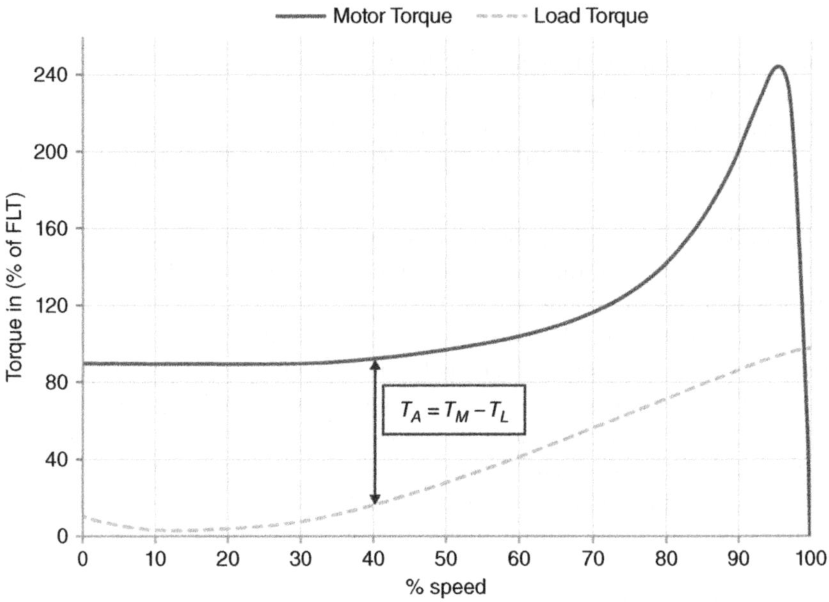

Figure 4.8 Motor acceleration torque. Source: IEEE Std 3002.7TM-2018 [7]/with permission of IEEE.

T_a is obtained,

$$T_a = T_m - T_l \tag{4.38}$$

The magnitude of T_a determines how fast the motor will accelerate as depicted in Figure 4.8.

Leveraging the general motion equation for rotating systems, it can be obtained that,

$$J\frac{d\omega}{dt} = T_a \quad \text{or} \quad \frac{1}{2g}WR^2\frac{d\omega}{dt} = T_a \tag{4.39}$$

where,

g is the acceleration due to gravity.
ω is the motor angular speed.

With the updated T_a, motor speed ω will be updated from Equation (4.39) which will be substituted into the equivalent circuit model Figure 3.9 to recalculate the equivalent impedance seen at the motor terminal. The recalculated impedance is used to modify the network admittance equation for solving the latest system bus voltages. This process continues until the motor is fully accelerated or a specified simulation time ends.

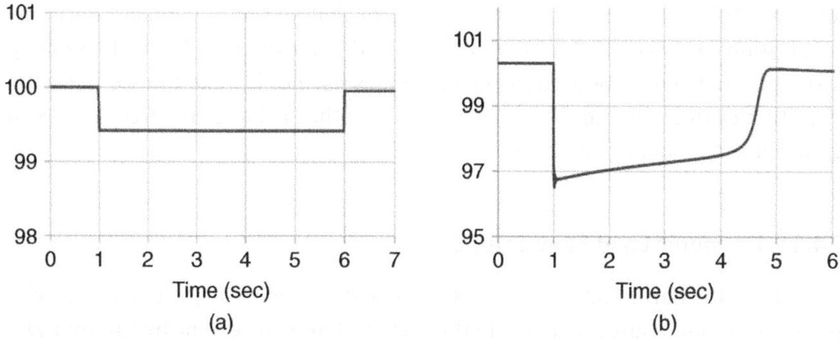

Figure 4.9 Motor voltage comparison for motor starting analysis. (a) Static motor starting and (b) dynamic motor starting. Source: IEEE Std 3002.7™-2018 [7]/with permission of IEEE.

Table 4.1 Component modeling comparison for motor starting analysis.

Component	Static solution	Dynamic solution
Generator	Constant voltage behind X'_d	Constant voltage behind X'_d
Utility	Constant voltage behind X_{sc}	Constant voltage behind X_{sc}
Operating motor	Constant power (kVA)	Constant power (kVA)
Starting motor	Locked-rotor Z with starting time	Dynamically modeled
Motor starting device	Modeled	Modeled
Voltage controller and regulator	Modeled	Modeled

Source: Adapted from IEEE Std 3002.7™-2018 [7].

The dynamic motor starting solution is carried in time domain. Different than the static solution, dynamic solution results show continuous curves of motor and system responses to provide great details of the motor performance and system voltage variations for in depth analysis. A comparison of motor voltages from the static solution and dynamic solution is shown in Figure 4.9. In addition, Table 4.1 shows differences in system component modeling between two solutions.

4.4 Harmonic Solution

Harmonic analysis is based on the system harmonic sources to provide information on harmonic voltage and current distortions, identify parallel and series harmonic resonance conditions, assist filter design and tuning, and offer other essential data for engineers to perform detailed and complete harmonic penetration investigation and mitigation. For the above tasks, basically there are two sets of harmonic solutions available, one is harmonic load flow for harmonic distortion evaluation and filter sizing, and the other is harmonic frequency scan for harmonic resonance assessment.

4.4.1 Harmonic Load Flow Solution

In applying the harmonic load flow solution, first, it is necessary to determine what are the harmonic sources existing in the system. This can be done by summarizing all harmonic sources in the study network and constructing a harmonic library $L(h)$ to list all harmonic sources:

$$\{L(h)\} = \{[I(h_1)], \dots, [I(h_{max})]\} \tag{4.40}$$

where,

$[I(h_j)] = [I_1(h_j), \dots, I_N(h_j)]$ is a harmonic current injection vector for h_j^{th} harmonic.
N is the total number of buses.
h_{max} is the highest order of the harmonics existing in the system.

Harmonic load flow solution is like the regular load flow solution with several fundamental modifications:

1) The admittance equation $Y_{bus}(h)$ is frequency (harmonic order) dependent; therefore, when analyzing a specific harmonic impact to the network, reactance, and admittance in the system component circuit model must be adjusted to the harmonic order per component harmonic models described in Chapter 3.
2) The admittance equation is also sequence dependent, therein, sequence of each harmonic whether it behaves like positive, negative, or zero sequence needs to be realized so that correct system topologies will be incorporated in building the $Y_{bus}(h)$.
3) There will be a number of admittance equations, each one corresponding to one of the existing harmonic orders specified in the harmonic library.

In general, harmonic admittance equation can be expressed as [8],

$$\begin{bmatrix} V_1(h) \\ V_2(h) \\ \vdots \\ V_N(h) \end{bmatrix} = \begin{bmatrix} Y_{11}(h) & Y_{12}(h) & \cdots & Y_{1N}(h) \\ Y_{21}(h) & Y_{22}(h) & \cdots & Y_{2N}(h) \\ \vdots & \vdots & \vdots & \vdots \\ Y_{N1}(h) & Y_{N2}(h) & \cdots & Y_{NN}(h) \end{bmatrix} \begin{bmatrix} I_1(h) \\ I_2(h) \\ \vdots \\ I_N(h) \end{bmatrix} \quad (h = h_1, \dots, h_{max}) \tag{4.41}$$

Iterate the harmonic library and solve each harmonic admittance Equation (4.41) for harmonic voltage vector $[V(h)] = [V_1(h), \dots, V_N(h)]$ for all harmonics $h = h_1, \dots, h_{max}$.

From each harmonic voltage vector, the harmonic current flows are calculated. For example, harmonic current from node i to node j through a branch with admittance $Y_{ij}(h)$ at the hth harmonic is,

$$I_{ij}(h) = Y_{ij}(h) \times (V_i(h) - V_j(h)) \tag{4.42}$$

After the node harmonic voltages and branch harmonic currents are solved, important harmonic distortion indices can be derived with the following equations [9]:

1) Total Demand Distortion (TDD) index:

For current $$TDD = 100 \times \frac{\sqrt{\sum_{h=2}^{\infty} I_h^2}}{I_{1,max}} \tag{4.43}$$

where h is the significant harmonic component order that usually goes up to 50, I_h is the hth harmonic current content and $I_{1,\,max}$ is the maximum demand load current at the fundamental frequency.

2) Total Harmonic Distortion (THD) index:

For voltage $$THD = 100 \times \frac{\sqrt{\sum_{h=2}^{\infty} V_h^2}}{V_1} \tag{4.44a}$$

For current $$THD = 100 \times \frac{\sqrt{\sum_{h=2}^{\infty} I_h^2}}{I_1} \tag{4.44b}$$

where h is the significant harmonic component order that usually goes up to 50, V_h, and I_h are the hth harmonic voltage and current content, V_1 and I_1 are the fundamental frequency voltage and current, respectively.

3) Individual Harmonic Distortion (IHD) index:

For voltage $$IHD = 100 \times \frac{V_h}{V_1} \tag{4.45a}$$

For current $$IHD = 100 \times \frac{I_h}{I_1} \tag{4.45b}$$

where h is harmonic component order that usually goes up to 50, V_h and I_h are the hth harmonic voltage and current content respectively.

4) RMS value:

For voltage $$V_{rms} = \sqrt{\sum_{h=1}^{\infty} V_h^2} \tag{4.46a}$$

For current $$I_{rms} = \sqrt{\sum_{h=1}^{\infty} I_h^2} \tag{4.46b}$$

where h is harmonic component order including a specific order of $h = 1$ that stands for the fundamental frequency component.

5) Telephone Interference Factor (TIF):

For voltage $$TIF = \frac{\sqrt{\sum_{h=1}^{\infty}(K_h \times P_h \times V_h)^2}}{\sqrt{\sum_{h=1}^{\infty} V_h{}^2}} \tag{4.47a}$$

For current $$TIF \frac{\sqrt{\sum_{h=1}^{\infty}(K_h \times P_h \times I_h)^2}}{\sqrt{\sum_{h=1}^{\infty} I_h{}^2}} \tag{4.47b}$$

where K_h and P_h are the weighting factors.

4.4.2 Harmonic Frequency Scan Solution

Frequency scan is to vary the harmonic injection frequency and scan impedance at each critical point to identify parallel resonance conditions in the network. In a parallel resonance case, the maximum bus voltage will be observed at the resonant buses; and in a series resonance case, the maximum current will be observed at the resonant branches. Both parallel and series resonance occurrences and locations depend on frequencies of the harmonic injection. The parallel resonance accounts for the majority of harmonic resonance problems in I&C power systems.

The solution of the harmonic frequency scan is based on the system impedance equation. With the assumption that only bus k has a harmonic current injection, the impedance equation can be written as,

$$\begin{bmatrix} V_1(h) \\ \vdots \\ V_k(h) \\ \vdots \\ V_N(h) \end{bmatrix} = \begin{bmatrix} Z_{11}(h) \cdots Z_{1k}(h) \cdots Z_{1N}(h) \\ \vdots \quad \ddots \quad \vdots \quad \ddots \quad \vdots \\ Z_{k1}(h) \cdots Z_{kk}(h) \cdots Z_{kN}(h) \\ \vdots \quad \ddots \quad \vdots \quad \ddots \quad \vdots \\ Z_{N1}(h) \cdots Z_{Nk}(h) \cdots Z_{NN}(h) \end{bmatrix} \begin{bmatrix} 0 \\ \vdots \\ I_k(h) \\ \vdots \\ 0 \end{bmatrix} \tag{4.48}$$

Voltages, therefore, at each bus are,

$$V_i(h) = Z_{ik}(h) \times I_k(h) \quad (i = 1, \ldots, \mathrm{N}) \tag{4.49}$$

where,

$Z_{ik}(h)$ $(i \neq k)$ is called transfer impedance.
$Z_{kk}(h)$ is called driving point impedance.

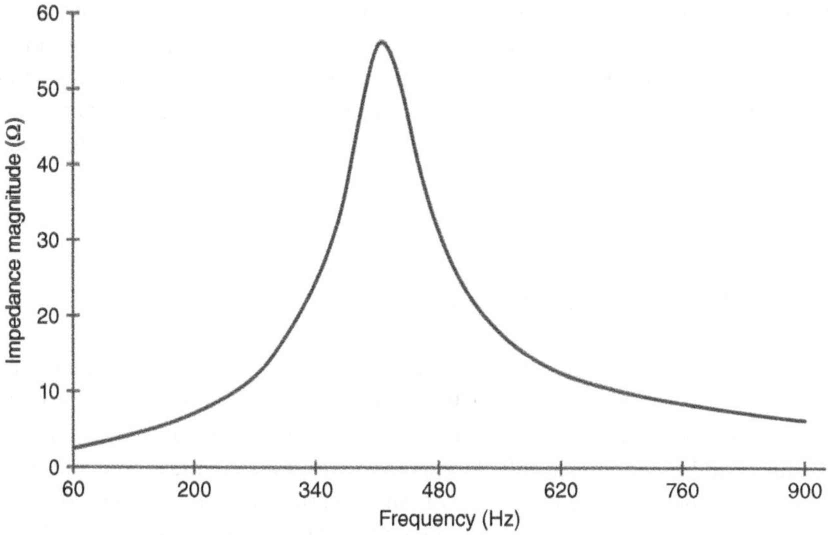

Figure 4.10 Impedance magnitude versus frequency for parallel resonance. Source: IEEE Std 3002.8™-2018 [9]/with permission of IEEE.

Both are the function of frequency (harmonic order). Value and shape of the driving point impedance $Z_{kk}(h)$ contain the information for parallel resonance. If a parallel resonance exists, at the resonant frequency $h = h_{res}$, magnitude $Z_{kk}(h_{res})$ reaches to the maximum (either local maximum or global maximum) as shown in Figure 4.10, and the phase angle of $Z_{kk}(h_{res})$ flips the sign. This is due to the factor that at the resonant frequency, the imaginary part of the bus driving point admittance $Y_{kk}(h_{res}) = 1/Z_{kk}(h_{res})$ becomes zero since the inductive component and capacitive component are canceled by each other, thus $Y_{kk}(h_{res})$ reaches to the minimum or $Z_{kk}(h_{res})$ becomes the maximum.

The peak value of $|Z_{kk}(h_{res})|$ could be tenth- or hundredth-times higher than the value under the fundamental frequency $h = 1$. When it is multiplied by the harmonic current at this frequency, it if exists, it would produce a voltage amplification effect which often results in sever over-voltage condition to damage the equipment connected to the bus, i.e.

$$|V_k(h_{res})| = |Z_{kk}(h_{res})| \times |I_k(h_{res})| \gg |V_k(1)| \tag{4.50}$$

The complete harmonic frequency scan starts from a lower frequency to a higher frequency by changing the frequency at every delta step and performing the above-described driving point impedance calculation at every location specified, in particular, where power factor correction capacitors or other reactive var compensation device installed. This is due to the fact that the I&C power systems are normally full of motors and are seen primarily as inductive load dominant network which is highly susceptible to parallel resonance if there are capacitive component connected in parallel. With all calculated driving point impedance data gathered, connecting those data will form a frequency scan curve as shown in Figure 4.10 for each specified location. Scanning locations, as well as the frequency scanning range are defined by the engineers who are performing the study and analysis.

4.5 Transient Stability Solution

Transient stability solution is to solve the combined machine dynamic equations and network algebra equations to obtain time domain system responses covering both machine and network variables. In particular, the control systems of machines and other controllable devices are also simulated and solved simultaneously together with the machine dynamic equations and the network algebra equations. The transient stability simulation involves a large number of variables, and has to deal with numerous high-order algebra and complex differential equations. It continues throughout the duration in simulation time range. To perform the transient simulation, an initial steady state solution is needed which also provides valuable insight to system state variable set points and reveals special information on the system status. Once the system is initialized, the remaining simulations in time domain constitute the full dynamic solution.

4.5.1 Initialization and Steady-State Solution

4.5.1.1 Data Preparation

Transient stability simulation begins from an initial solution or initialization. This is a vital stage to allow a successful and reliable transient stability solution. During this stage, a base case of system operating condition defined by network configuration, generator power output, utility connection, load profile, as well as operating modes and settings for all control systems must be specified. Detailed data preparation requirements are listed below.

1) Specify system configurations including all switching device status in either open or closed position.
2) Confirm utility inter-ties whether they are connected or disconnected.
3) Decide generator control modes for voltage control, active and reactive power control, and speed control.
4) Dispatch generator active power and reactive power generation according to the control modes and desired generation schedule.

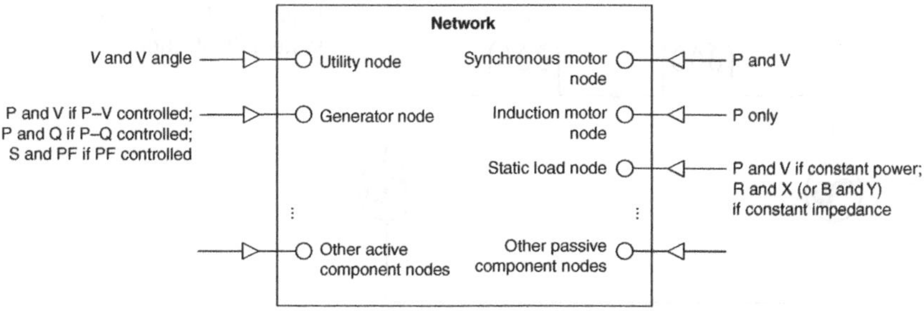

Figure 4.11 Illustration of initial conditions for various components.

5) Based on the operating condition and requirement, choose fixed set points for all generator controls, including excitation and AVR, speed governor, and power system stabilizer (PSS).
6) Assign loading for each motor load according to their operating conditions and the study scenario chosen. Reactive power loading of induction motor normally cannot be preassigned since they depend on the connected system and the actual terminal voltages. Therefore, the induction motor reactive power will be solved during the initialization.
7) Determine other loads with their active power and reactive power loading or equivalent impedance per their operating schedule and characteristics.
8) Review and acknowledge all other control devices for their modes and fixed settings.

Figure 4.11 illustrates examples of initial conditions for different components.

4.5.1.2 Initialization

Once all the data and information are provided, the transient stability simulation will perform a steady state solution which is also referred to as initialization of the system to solve for all state variables for the entire activated system, which include,

1) Network state variables, i.e. bus voltages.
2) Machine state variables inside all operating machine equations, including synchronous machine *d–q* models and induction machine equivalent circuit models.
3) Control state variables of supplement active control systems, including excitation and AVR, speed governor, PSS, SVC, and all other dynamic control devices.

4.5.1.3 Steady-State Stability

One important piece of information from the initialization solution for synchronous generator is its equivalent internal bus voltage magnitude and phase

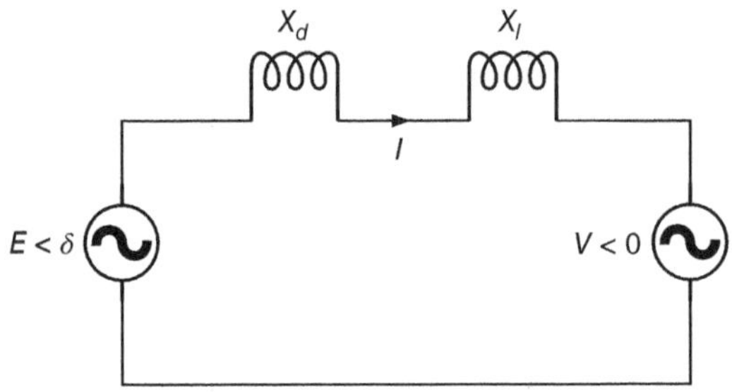

Figure 4.12 Two-machine system illustration.

angle, which determine the steady state stability for the generator. It can be proven that the electrical rotor angle of a synchronous generator is equivalent to the phase angle of its internal voltage. Figure 4.12 is a two-machine equivalent for a synchronous generator with respect to a slack bus (infinite of reference bus) in a simplified model. The slack bus is represented by a reference voltage source with magnitude V and phase angle 0°, and the generator is represented by a voltage source with a magnitude E and phase angle δ which is also the generator rotor angle. Two voltage sources are connected through the generator equivalent impedance X_d and a network impedance X_l.

The active (electrical) power P_e transferred from the generator to the network is a function of the rotor angle as shown in Eq. (4.51).

$$P_e = \frac{EV}{(X_d + X_l)} \sin\delta = P_{max} \sin\delta \tag{4.51}$$

where,

$$P_{max} = \frac{EV}{(X_d + X_l)} \tag{4.52}$$

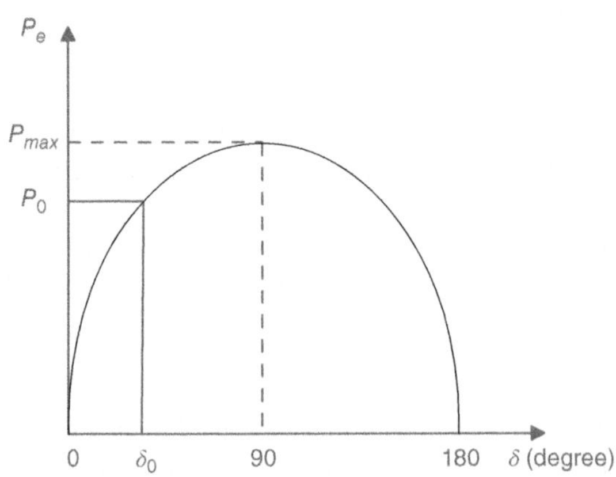

Figure 4.13 Power transfer capability curve.

Equation (4.52) can be plotted into the power transfer capability curve shown in Figure 4.13, where the generator power P_e that can be transferred from the generator into the network is subject to small disturbances and thus is called the steady-state stability limit.

P_e consists of two components, namely, synchronizing power and damping power as shown in the first and the second parts on the right side of Eq. (4.53.):

$$\Delta P_e = K_s \Delta \delta + K_d \Delta \omega \tag{4.53}$$

where the synchronizing power (or torque) component is in phase with the rotor angle deviation, and the damping power (or torque) component is in phase with the speed deviation. Lack of sufficient synchronizing power will result in aperiodic instability, whereas lack of damping power will result in oscillatory instability.

The synchronizing power coefficient k_s is found form Equation (4.51):

$$K_s = \frac{EV}{(X_d + X_l)} \cos \delta \tag{4.54}$$

and damping power coefficient K_d is the machine damping constant D:

$$K_d = D\,(= 0 \text{ } if \text{ } D = 0) \tag{4.55}$$

Since the synchronizing power must be greater than 0, from Equation (4.54), the steady state stability requirement for a synchronous machine can be defined by the region $[0, 90^\circ]$ for rotor angle δ to ensure a positive K_s, i.e.:

$$0 < \delta < 90^\circ \tag{4.56}$$

Thus, from the initialization of the system for transient stability solution, synchronous generator equivalent internal voltages and their phase angles will be found which are equivalent to the generator rotor angle. By Equation (4.56.), the steady state stability margins for generators can be evaluated.

4.5.2 Dynamic Solution

4.5.2.1 Main Iteration and Solution Process

After the system initial conditions are found and all state variables are initialized for time $t = 0$, dynamic simulation begins to alternatively solve the network equation, machine equations together with the machine and other active control equations aggressively in Δt time increment until the specified final time is reached. Suppose an I&C power system has N nodes with nodes 1 to m are dynamically modeled machines and active control buses and nodes $m + 1$ to N are load buses. The steps of dynamic solution can be summarized below.

1) Solve the generator excitation and AVR system equations to obtain the excitation voltage E_{fd} for each synchronous machines as explained in Chapter 3

$$\dot{x} = Ax + Bu$$

$$E_{fd} = Cx$$

2) From the synchronous machines' models and equations, derive the equivalent internal voltage E_{eq} and impedance of the machine Z_{eq} for each machine.
3) Use E_{eq} and Z_{eq} to construct a Thevenin equivalent for each synchronous machine.
4) For all other induction machines and active controllers, derive Thevenin equivalent E_{eq} and Z_{eq} from their models and parameters. For all passive controllers, derive their equivalent impedance and merge them into the admittance matrix Y_{bus}.
5) Covert all loads into shunt admittance and incorporate them into Y_{bus}. For constant power loads, following equations are applied to obtain the equivalent load admittance Y_{eq} where the load bus $V^{(0)}$ is from the initial voltage vector. Bus voltage vector will be updated in every time iteration to derive new Y_{eq} which will be applied to a new Y_{bus} to solve a new network equation.

$$Y_{eq} = \frac{P}{|V^{(0)}|^2} - j\frac{Q}{|V^{(0)}|^2} \tag{4.57}$$

6) The terminal representation of the system can then be explicitly expressed as Figure 4.14. The corresponding bus admittance equation for Figure 4.14 can be written in Equation (4.58), where $y_i = 1/Z_{eq,i}$ and $E_i = E_{eq,i}$ for machine buses.

$$\begin{bmatrix} y_1E_1 \\ \vdots \\ y_mE_m \\ 0 \\ \vdots \\ 0 \end{bmatrix} = \begin{bmatrix} (Y_{11}+y_1)\cdots & Y_{1m} & Y_{1,m+1} & \cdots & Y_{1N} \\ \vdots & \ddots & \vdots\ \vdots & \ddots & \vdots \\ Y_{m1} & \cdots(Y_{mm}+y_m) & Y_{m,m+1} & \cdots & Y_{mN} \\ Y_{m+1,1} & \cdots\ Y_{m+1,m} & Y_{m+1,m+1} & \cdots & Y_{m+1,N} \\ \vdots & \ddots & \vdots\ \vdots & \vdots & \vdots \\ Y_{N1} & \cdots\ Y_{Nm} & Y_{N,m+1} & \ddots & y_{NN} \end{bmatrix} \begin{bmatrix} V_1 \\ \vdots \\ V_m \\ V_{m+1} \\ \vdots \\ V_N \end{bmatrix} \tag{4.58}$$

7) Solve all bus voltages from Equation (4.58).
8) Solve each machine and active control current I_i from external bus voltage V_i and machine and controller internal voltage E_i using the equivalent impedance $Z_{eq,\,i}$,

$$I_i = \frac{E_i - V_i}{Z_{eq,i}} \qquad (i = 1, \ldots, m) \tag{4.59}$$

9) Compute machine electrical torque P_e at the internal voltage terminal which is the electrical power to balance with the machine mechanical power on the

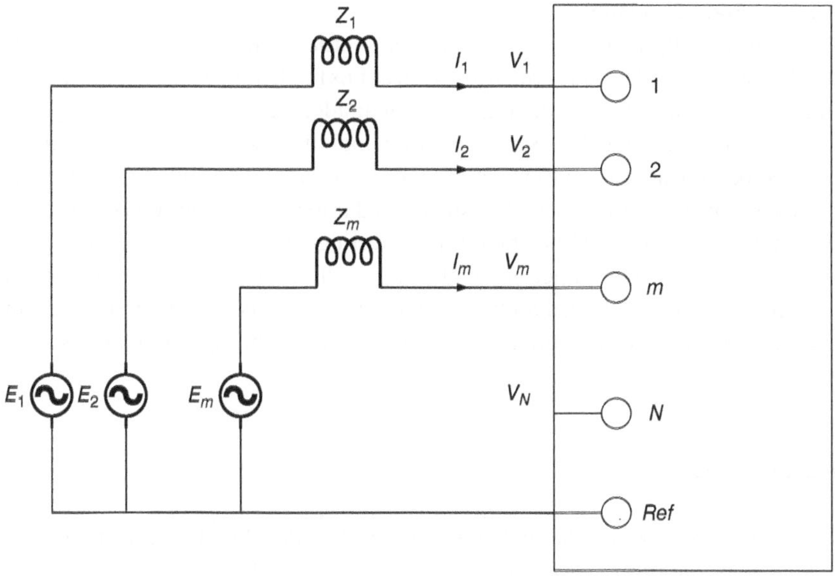

Figure 4.14 Terminal representation augmented by machine and active controller models.

shaft,

$$P_{e,i} = Re\left(E_i I_i^*\right) \qquad (i = 1, \ldots, m) \tag{4.60}$$

10) For each generator, compute shaft mechanical power $P_{m,\,i}$ from the speed governor control state space equation as explained in Chapter 3,

$$\dot{x} = A_i x + B_i u$$

$$P_{m,i} = C_i x \qquad (i = 1, \ldots, m \textit{ where i is a generator bus})$$

For motor, obtain P_m from the motor torque–slip curve.

11) Solve each machine speed from the swing equation as described in Chapter 3,

$$M_i \frac{d\omega_i}{dt} = P_{m,i} - P_{e,i}$$

Solve synchronous machine rotor angle for each generator from the question explained in Chapter 3,

$$\delta_i = \frac{d\omega_i}{dt} - \delta_{0,i}$$

Steps 1 to 11 constitute one complete transient stability dynamic solution cycle. When all variables are solved from the initial condition $t = 0$, the calculation will move to the next time step at Δt and go on. The process will continue and repeat progressively until the desired time range is filled with point-by-point solutions.

4.5.2.2 Event Simulation

Transient stability analysis is an event driven solution. Various time related events will be assigned to the study scenarios, such as bus fault, fault isolation, equipment outage or disconnection, sudden generation and load change, relay tripping, etc. The specified events will be inserted into the simulation loop and appropriate models for process to adjust equation parameters. For example, if a fault is placed at bus i, then a constraint of $V_i = 0$ will be substituted into the bus admittance equation. For another example, if a line or circuit connecting bus i and bus j is opened, then the corresponding admittance element in Y_{bus} will be set to 0, i.e. $Y_{ij} = 0$ (parallel connected circuits will need additional handling). Events are either specified in the study scenarios or triggered by the control actions through simulations such as replys.

References

1 Ferreira, C.A. (2005). A Second Order Power Flow Based on Current Injection Equations. *Electrical Power & Energy Systems* 27 (4): 254–263.

2 IEEE 3002.3–2018, IEEE Recommended Practice for Conducting Short-Circuit Studies and Analysis of Industrial and Commercial Power Systems, IEEE, 2018.

3 IEEE Std 399-1997, IEEE Recommended Practice for Industrial and Commercial Power Systems Analysis, IEEE, 1997.

4 IEEE Std 551™-2006, IEEE Recommended Practice for Calculating Short-Circuit Currents in Industrial and Commercial Power Systems, IEEE, 2006.

5 ANSI/IEEE C37.5-1979, IEEE Guide for Calculation of Fault Currents for Application of AC High-Voltage Circuit Breakers Rated on a Total Current Basis, IEEE, 1979.

6 IEEE Std C37.101-2016, IEEE Application Guide for AC High Voltage Circuit Breakers > 1000 Vac Rated on a Symmetrical Current Basis, IEEE, 2016.

7 IEEE Std 3002.7™-2018, Recommended Practice for Conducting Motor-Starting Studies and Analysis of Industrial and Commercial Power Systems, IEEE, 2018.

8 IEEE Power Engineering Society, *Tutorial on Harmonics Modeling and Simulation*, Publication No. 98TP125-0, IEEE, 1998.

9 IEEE Std 3002.8™-2018, IEEE Recommended Practice for Conducting Harmonic Studies and Analysis of Industrial and Commercial Power Systems, IEEE, 2018.

5

Numerical Simulation Techniques

Numerical simulation techniques play a crucial role in industrial and commercial (I&C) power system analysis because computer tools heavily rely on them to solve various equations describing system static and dynamic behaviors and simulate normal and abnormal operation conditions. Numerical techniques involve discretizing the continuous time domain functions using time step. At each step instance, the equations are solved and the time is advanced to the next step. The equations, which have been discussed in the previous chapters, include network admittance and impedance equations in different formats, differential equations for machine and control systems, and other dynamic equipment and devices, as well as the overall differential algebraic equation (DAE) for the entire network for time domain analysis. The matured numerical computation methods are implemented in computer software for general-purpose applications. Understanding fundamentals of some of the applicable numerical simulation methods and their typical applications in I&C power system analysis provides background information and assistance for data preparation and validation, study parameter selection, and control of solution feasibility and accuracy which are all imperative to successful system analysis.

5.1 Linear Algebraic Equation

A linear algebraic equation describes a relationship between n real or complex number variables $x_1, x_2, \ldots, x_n$ in a linear format, i.e.

$$a_1x_1 + a_2x_2 + \cdots + a_nx_n = b_1 \tag{5.1}$$

where $a_1, a_2, \ldots, a_n$ and b_1 are all real or complex number constant coefficients.

Industrial and Commercial Power System Analysis Fundamentals and Practice, First Edition. J. J. Dai.

If there are m of independent relationships that can be established,

$$\begin{array}{llll} a_{11}x_1+ a_{12}x_2 + & \dots & +a_{1n}x_n = b_1 \\ a_{21}x_1+ a_{22}x_2 + & \dots & +a_{2n}x_n = b_2 \\ \dots \quad \dots & \dots & \dots \\ a_{m1}x_1+ a_{m2}x_2 + & \dots & a_{mn}x_n = b_m \end{array} \tag{5.2}$$

that forms a linear algebraic equation set.

In particular, when the number of equations m is the same as the number of variables n, there is,

$$\begin{array}{llll} a_{11}x_1+ a_{12}x_2 + & \dots & +a_{1n}x_n = b_1 \\ a_{21}x_1+ a_{22}x_2 + & \dots & +a_{2n}x_n = b_2 \\ \dots \quad \dots & \dots & \dots \\ a_{n1}x_1+ a_{n2}x_2 + & \dots & +a_{nn}x_n = b_n \end{array} \tag{5.3}$$

Or written it in matrix form,

$$\boldsymbol{Ax} = \boldsymbol{b} \tag{5.4}$$

where

$\boldsymbol{A} = \begin{bmatrix} a_{11} & a_{12} & \dots & a_{1n} \\ a_{21} & a_{22} & \dots & a_{2n} \\ \vdots & \vdots & \ddots & \vdots \\ a_{n1} & a_{n2} & \dots & a_{nn} \end{bmatrix}$ is the coefficient matrix and is square,

$\boldsymbol{b} = \begin{bmatrix} b_1 & b_2 & \dots & b_n \end{bmatrix}$ is the right-hand side vector,

$\boldsymbol{x} = \begin{bmatrix} x_1 & x_2 & \dots & x_n \end{bmatrix}$ is the variable vector.

In I&C power systems, Equation (5.4) can represent the nodal admittance equation or nodal impedance equation,

$$Y_{bus}V = I \tag{5.5}$$

$$Z_{bus}I = V \tag{5.6}$$

The number of variables equals the number of buses.

Mathematically, it can be proven that if all equations in Equation (5.4) are linearly independent, i.e. if $rank(A) = n$, then the inverse of a matrix $\boldsymbol{A}^{-1}$ exists theoretically, and the linear system (5.4) has a unique solution,

$$\boldsymbol{x} = \boldsymbol{A}^{-1}\boldsymbol{b} \tag{5.7}$$

Equation (5.7) can be satisfied in general by I&C power systems.

There are two types of methods to solve linear algebraic equations, direct method and indicate method. The direct methods are based on elimination approaches to reduce variables to the level that a simple solution for one variable is found. Then based on the solved variable(s), substitution process is carried out to solve remaining variables one by one, each step involves only simple manipulations. The indirect methods, which often depend on iterative process, solve variables in an approximation fashion and repeat the process sequentially, each step improves the solution precision.

5.1.1 Direct Solution Method

5.1.1.1 Gaussian Elimination

Based on Equation (5.3), this method starts by eliminating the variable x_1 from the second to the last equations, then eliminating x_2 from the third to the last equations and continues until only x_n left in the last equation. The final modified Equation (5.3) is an upper-triangular set of equations given by,

$$\begin{array}{llll} a_{11}^{(1)}x_1 + & a_{12}^{(1)}x_2 + & \dots & +a_{1n}^{(1)}x_n = b_1^{(1)} \\ 0+ & a_{22}^{(2)}x_2 + & \dots & +a_{2n}^{(2)}x_n = b_2^{(2)} \\ \dots & \dots & \dots & \dots \\ 0+ & 0+ & \dots & +a_{nn}^{(n)}x_n = b_n^{(n)} \end{array} \tag{5.8}$$

Where the subscripts stand for the step of update. The elimination steps are shown below,

1) For n equations (from the 1st to the nth equations), perform,

$$a_{ij}^{(1)} = a_{ij}, i, j = 1, 2, \dots, n \tag{5.9a}$$

$$b_i^{(1)} = b_i, i = 1, 2, \dots, n \tag{5.9b}$$

2) After the above is done, for the last $n - 1$ equations (from the 2nd to the nth equations), perform,

$$m_{i1} = \frac{a_{i1}^{(1)}}{a_{11}^{(1)}}, i = 2, 3, \dots, n \tag{5.9c}$$

$$a_{ij}^{(2)} = a_{ij}^{(1)} - m_{i1}a_{1j}^{(1)}, \quad i = 2, 3, \dots, n, j = 1, 2, \dots, n \tag{5.9d}$$

$$b_i^{(2)} = b_i^{(1)} - m_{i1}b_i^{(1)}, \quad i = 2, 3, \dots, n \tag{5.9e}$$

3) Repeat step (2) with i moves incrementally at each step until the last equation (the nth equation). At each step for x_k is to be eliminated, perform,

$$m_{ik} = \frac{a_{ik}^{(k)}}{a_{kk}^{(k)}}, \quad i = k + 1, k + 2, \dots, n \tag{5.10a}$$

$$a_{ij}^{(k+1)} = a_{ij}^{(k)} - m_{ik}a_{kj}^{(k)}, \quad i = k + 1, k + 2, \dots, n, j = k, k + 1, \dots, n \tag{5.10b}$$

$$b_i^{(k+1)} = b_i^{(k)} - m_{ik}b_k^{(k)}, \quad i = k + 1, k + 2, \dots, n \tag{5.10c}$$

Finally, Equation (5.8) is obtained.

Equation (5.8) is in a triangular form and can be easily solved by a back-substituting, starting from the last equation (the nth equation) to the 1st equation. All variables from x_n to x_1 are calculated subsequently from,

$$x_n = \frac{b_n^{(n)}}{a_{nn}^{(n)}} \tag{5.11a}$$

$$x_{n-1} = \frac{b_{n-1}^{(n-1)} - a_{n-1n}^{(n-1)} x_n}{a_{n-1n-1}^{(n-1)}} \tag{5.11b}$$

$$\cdots \quad \cdots \quad \cdots \quad \cdots$$

$$x_k = \frac{b_k^{(k)} - a_{kk+1}^{(k)} x_{k+1} - \ldots - a_{kn}^{(k)} x_n}{a_{kk}^{(k)}}, \qquad k = n-2, \ldots, 1 \tag{5.11c}$$

5.1.1.2 LU (Lower-Upper) Decomposition

This method attempts to decompose the coefficient matrix $\boldsymbol{A}$ into an upper-triangular matrix $\boldsymbol{U}$ and a lower-triangular matrix $\boldsymbol{L}$, such that,

$$\boldsymbol{A} = \boldsymbol{LU} \tag{5.12}$$

where

$$L = \begin{bmatrix} l_{11} & 0 & \cdots & 0 \\ l_{21} & l_{11} & \cdots & 0 \\ \vdots & \vdots & \ddots & \vdots \\ l_{n1} & l_{n2} & \cdots & l_{nn} \end{bmatrix} \tag{5.13}$$

$$U = \begin{bmatrix} 1 & u_{12} & \cdots & u_{1n} \\ 0 & 1 & \cdots & u_{2n} \\ \vdots & \vdots & \ddots & \vdots \\ 0 & 0 & \cdots & 1 \end{bmatrix} \tag{5.14}$$

Coefficients of $\boldsymbol{U}$ and $\boldsymbol{L}$ can be computed systemically. The linear algebraic Equation (5.4) can then be written as,

$$\boldsymbol{LUx} = \boldsymbol{b}$$

By introducing an intermittent variable vector $\boldsymbol{y}$,

$$\boldsymbol{Ux} = \boldsymbol{y} \tag{5.15}$$

such that,

$$\boldsymbol{LUx} = \boldsymbol{b} \Rightarrow \boldsymbol{Ly} = \boldsymbol{b} \tag{5.16}$$

Table 5.1 Comparison of operational counts [1].

Method	Operational counts
Gaussian elimination	$\frac{n^2}{3}+n^2-\frac{n}{3}$
Triangular decomposition	$\frac{n^2}{3}+n^2-\frac{n}{3}$

Source: Williams [1]/with permission of Thomas Nelson & Sons Ltd.

Then Equation (5.4) can be solved in two steps, with the first step to perform a forward-substituting on Equation (5.16) to calculate $\boldsymbol{y}$, and the second step is to perform a backward-substituting on Equation (5.15) to calculate $\boldsymbol{x}$. The substitution processes are similar to the Gaussian elimination method.

Gaussian elimination and LU decomposition carry the same computation burden with about the same magnitude of operational counts, as shown in Table 5.1.

The direct solution can be applied to short-circuit, harmonic load flow and transient stability calculations.

5.1.2 Iterative Solution Method

5.1.2.1 Gauss–Seidel Method

Gauss–Seidel method first rewrites Equation (5.4) in the following formats,

$$\boldsymbol{A} = L + D + U \tag{5.17}$$

where

$$L = \begin{bmatrix} 0 & 0 & \dots & 0 \\ a_{21} & 0 & \dots & 0 \\ \vdots & \vdots & \ddots & \vdots \\ a_{n1} & a_{n2} & \dots & 0 \end{bmatrix}$$

$$U = \begin{bmatrix} 0 & a_{12} & \dots & a_{1n} \\ 0 & 0 & \dots & a_{2n} \\ \vdots & \vdots & \ddots & \vdots \\ 0 & 0 & \dots & 0 \end{bmatrix}$$

$$D = \begin{bmatrix} a_{11} & 0 & \dots & 0 \\ 0 & a_{22} & \dots & 0 \\ \vdots & \vdots & \ddots & \vdots \\ 0 & 0 & \dots & a_{nn} \end{bmatrix}$$

An iterative solution process for $\boldsymbol{x}$ can be formulated as,

$$[\boldsymbol{L}+\boldsymbol{D}+\boldsymbol{U}]\boldsymbol{x}=\boldsymbol{b} \tag{5.18}$$

or

$$\boldsymbol{x}=\boldsymbol{D}^{-1}\{\boldsymbol{b}-(\boldsymbol{L}+\boldsymbol{U})\boldsymbol{x}\} \tag{5.19}$$

Or create an iteration process and write for each x explicitly,

$$x_i^{r+1}=\frac{1}{a_{ii}}\left\{b_i-\sum_{j\neq i}^{n}a_{ij}\,x_j^r\right\},\quad i=1,2,\ldots,n \tag{5.20}$$

where r is the iteration number. By adding an acceleration factor α, the accelerated Gauss–Seidel method can be formulated to improve the convergence rate,

$$x_i^{r+1}=\frac{\alpha}{a_{ii}}\left\{b_i-\sum_{j\neq i}^{n}a_{ij}\,x_j^r\right\}-(\alpha-1)x_i^r,\quad i=1,2,\cdots,n \tag{5.21}$$

where

$$0<\alpha\leq 2 \tag{5.22}$$

The iteration starts from $r=1$ and continues until convergence criteria for a specified tolerance ε is met,

$$\left|x_i^{r+1}-x_i^r\right|<\varepsilon,\ \ for\ all\ \ i=1,2,\cdots,n \tag{5.23}$$

5.1.2.2 Newton–Raphson Method

Write Equation (5.4) as,

$$f(\boldsymbol{x})=\boldsymbol{A}\boldsymbol{x}=\boldsymbol{b} \tag{5.24}$$

By applying the Taylor series expansion on $f(\boldsymbol{x})$ about the true solution vector $\overline{\boldsymbol{x}}$,

$$f(\boldsymbol{x}-\overline{\boldsymbol{x}})\approx f(\overline{\boldsymbol{x}})+\frac{\partial f}{\partial \boldsymbol{x}}(\boldsymbol{x}-\overline{\boldsymbol{x}}) \tag{5.25}$$

where $\frac{\partial f}{\partial \boldsymbol{x}}$ is the first derivative of $f(\boldsymbol{x})$ and is a square matrix called the Jacobian $\boldsymbol{J}$,

$$J(x)=\frac{\partial f}{\partial \boldsymbol{x}}=\begin{bmatrix}\frac{\partial f_1}{\partial \boldsymbol{x}_1} & \frac{\partial f_1}{\partial \boldsymbol{x}_2} & \cdots & \frac{\partial f_1}{\partial \boldsymbol{x}_n}\\ \frac{\partial f_2}{\partial \boldsymbol{x}_1} & \frac{\partial f_2}{\partial \boldsymbol{x}_2} & \cdots & \frac{\partial f_2}{\partial \boldsymbol{x}_n}\\ \vdots & \vdots & \ddots & \vdots\\ \frac{\partial f_n}{\partial \boldsymbol{x}_1} & \frac{\partial f_n}{\partial \boldsymbol{x}_2} & \cdots & \frac{\partial f_n}{\partial \boldsymbol{x}_n}\end{bmatrix} \tag{5.26}$$

Substitute Equation (5.25) to Equation (5.24) and reorganize it, Newton–Raphson method formulates a solution to solve Equation (5.22) using successive evaluations of the Jacobian matrix and continuous estimation of $\boldsymbol{x}^r$ where r is the rth iteration.

$$0 = f(\overline{\boldsymbol{x}}) - \boldsymbol{b} \approx f(\boldsymbol{x}^r) - \boldsymbol{b} + J(\boldsymbol{x}^r)(\boldsymbol{x}^{r+1} - \boldsymbol{x}^r) \tag{5.27}$$

i.e.,

$$\boldsymbol{x}^{r+1} = \boldsymbol{x}^r + [J(\boldsymbol{x}^r)]^{-1}[b - f(\boldsymbol{x}^r)] \tag{5.28}$$

The convergence criteria are the same as in Equation (5.23) specified by the user.

The Newton–Raphson method takes the first derivative of the function to be solved so it finds a steeper direction for moving $\boldsymbol{x}^r$ to $\boldsymbol{x}^{r+1}$ and thus has a faster convergence rate. However, The Newton–Raphson method is not globally convergent, meaning that it does not always converge to a solution from every initial iteration set $\boldsymbol{x}^1$; therefore, more careful selection and additional effort might be needed to choose the initial guess. The iterative solution method is applied to solve I&C power system load flow equations.

5.2 Differential Equation

Differential equations exist extensively in I&C power system models, from machine rotor-stator impedance equations to generator excitation and governor control systems to other dynamic devices. These equations describe equipment and device characteristics and responses in transient and form part of system overall equations for dynamic motor starting and transient stability simulation and analysis. Examples of these equations of these differential equations are discussed in Chapter 3, for instance, the machine circuit model in Equation (3.2),

$$\left[\frac{dI}{dt}\right] = [L_2]^{-1}\{-[R] - \omega[L_1]\}[I] + [L_2]^{-1}[V]$$

And the machine swing equation Equation (3.3),

$$J\frac{d\omega}{dt} = T_m - T_e - D\omega$$

Numerical methods provide tools to solve these equations and are implemented in computer software calculations engines.

5.2.1.1 Euler's Method

The Euler method provides numerical solutions to the initial-value problems of ordinary differential equations. Consider a general form for the initial-value problem of a first-order differential equation,

$$\frac{dy}{dx} = f(x, y) \tag{5.29}$$

$$y(x_0) = y_0 \text{ as the initial step value} \tag{5.30}$$

The first-order Euler method is used to find the next step value explicitly by,

$$y_{n+1} = y_n + \Delta x f(x_n, y_n) \quad n = 0, 1, \ldots \tag{5.31}$$

where Δx is the step size.

Essentially, this approximates,

$$\frac{dy}{dx} \cong \frac{\Delta y}{\Delta x} = \frac{y_{n+1} - y_n}{\Delta x} \tag{5.32}$$

i.e., using a difference to replace the derivative. Figure 5.1 illustrates the principle of this method.

Due to the first-order approximation, the error of the Euler's method is proportional to the square of the step size.

5.2.1.2 Prediction–Correction Method

The prediction–correction method combines Euler's method and the trapezoidal rule to provide a second-order, explicit numerical solution to the differential equation initial-value problems. The trapezoidal rule to solve Equation (5.29) is given as,

$$y_{n+1} = y_n + \frac{1}{2}\Delta x[f(x_n, y_n) + f(x_{n+1}, y_{n+1})] \tag{5.33}$$

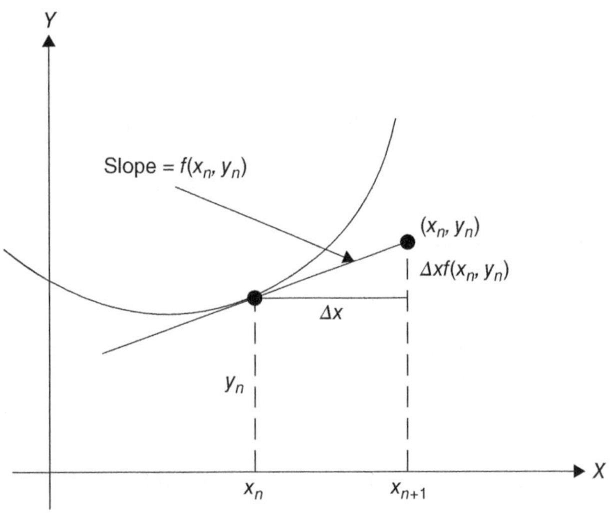

Figure 5.1 Euler's method.

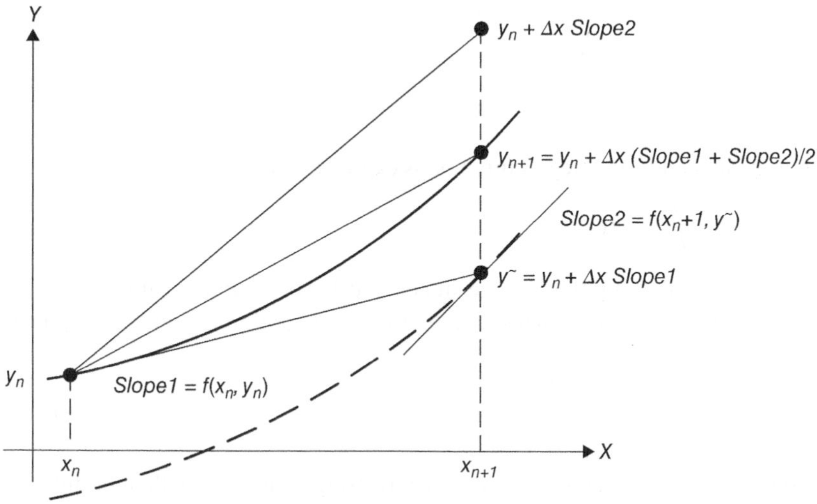

Figure 5.2 Prediction–correction method.

Notice Equation (5.33) is implicit, meaning the estimated value y_{n+1} appears on both sides of the equation. If applying Euler's method to estimate y_{n+1}, the prediction–correction method is formed. The prediction–correction process includes two steps, as explained below and in Figure 5.2,

Prediction of the next step value $\tilde{y}_{n+1}$,

a) $$\tilde{y}_{n+1} = y_n + \Delta x f(x_n, y_n) \quad n = 0, 1, \ldots \tag{5.34}$$

Correction and estimation of the final next step value y_{n+1},

b) $$y_{n+1} = y_n + \frac{1}{2}\Delta x[f(x_n, y_n) + f(x_{n+1}, \tilde{y}_{n+1})] \tag{5.35}$$

The prediction–correction method is a second-order solution and has generally higher precision than Euler's method.

The numerical solution methods for differential equations are applied to solve dynamic motor starting and transient stability simulations which involve machine and control system differential equations.

5.3 Integral Equation

Integral equations appear in functions of control devices of I&C power systems, such as an integrator in a governor or excitation systems demonstrated in Figure 5.3,

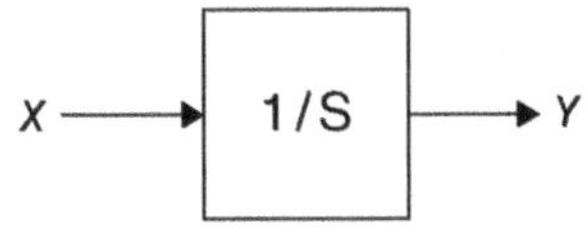

Figure 5.3 Integral function.

Mathematically, the integral function is expressed as,

$$y(x) = \int f(x)dx \tag{5.36}$$

For numerical computation, the upper and lower bounds of integration are known. Thus the value y corresponding to the integration region $[x_n, x_{n+1}]$ can be expressed as,

$$y = \int_{x_n}^{x_{n+1}} f(x)dx \tag{5.37}$$

Equation (5.37) can be solved numerically using Simpson's method or differential equation method.

5.3.1.1 Simpson's Method

Let $\Delta x = (x_{n+1} - x_n)/2$, Simpson's method approximate function $f(x)$ using a quadratic curve $Q(x)$ as illustrated in Figure 5.4 together with the trapezoidal rule so that Equation (5.37) can be estimated as,

$$y \approx \frac{\Delta x}{3}[f(x_n) + 4f(x_n + \Delta x) + f(x_n + 2\Delta x)] \tag{5.38}$$

Essentially this is equivalent to a numerical differential approach.

5.3.1.2 Differential Equation Method

Evaluating a definite integral value can be converted to the initial value problem for a differential equation. Consider,

$$y(x) = \int_a^x f(u)du \tag{5.39}$$

Per the fundamental theorem of calculus, by taking the differentiation of both sides of Equation (5.39), it can be written as,

$$\frac{dy}{dx} = f(x), y(a) = 0 \tag{5.40}$$

Equation (5.40) can then be solved by the trapezoidal rule for differential equations without the need of applying the Euler's method as it is an explicit function.

Control systems such as excitation and automatic voltage regulator, prime mover and governor, and variable frequency drive consist of integrations and need to apply the described methods here to solve.

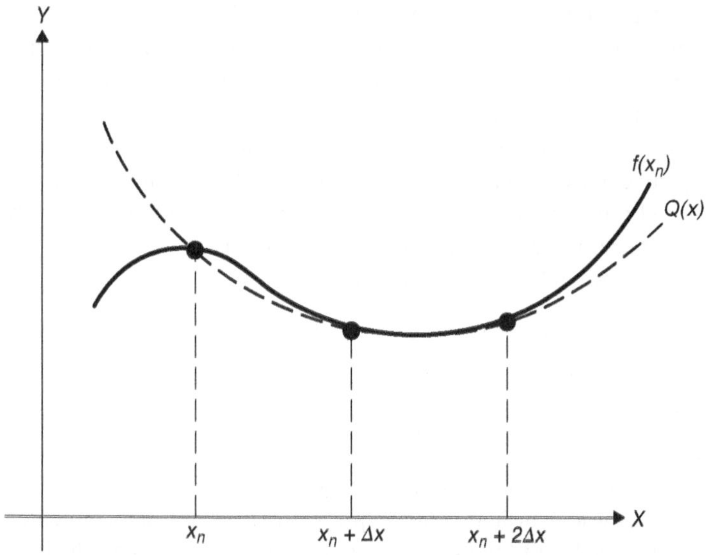

Figure 5.4 Simpson's method [1]. Source: Williams [1]/with permission of Thomas Nelson & Sons Ltd.

5.4 Mixed Linear Algebraic and Differential Equation

Recall the DAE mentioned in Chapter 3 and rewrite the general power system dynamics equations in combination with network equations in the compact form [2],

$$\frac{d\boldsymbol{x}}{dt} = F(t, \boldsymbol{x}, \boldsymbol{y}) \tag{5.41}$$

$$0 = G(t, \boldsymbol{x}, \boldsymbol{y}) \tag{5.42}$$

where,

$\boldsymbol{x}$ are the the state variables of the machine dynamics equations, and control device equations, including rotor angle, machine speed, transient and subtransient voltages on d-axis and q-axis of machine circuits, etc.

$\boldsymbol{y}$ are the algebraic variables for the network equations, including nodal voltages and nodal injection currents.

$\dot{\boldsymbol{x}}$ are the derivatives of $\boldsymbol{x}$.

$F(t, \boldsymbol{x}, \boldsymbol{y})$ are the state space equations describing the dynamics of machines and control devices.

$G(t, \boldsymbol{x}, \boldsymbol{y})$ are the network load flow equations.

Equations (5.41) and (5.42) can be solved by direct discretization methods. As one example, when the backward Euler method is applied to Equation (5.41), the steps are shown below [3],

1) Initialize the state variables $\boldsymbol{x}$ based on the given operating conditions (boundary conditions) to $\boldsymbol{x}_0$. In I&C power systems, these boundary conditions include generator terminal voltage and active power output, synchronous motor terminal voltage and shaft load, induction motor shaft load, etc. There are a few exceptions, such as generators in P–Q (active power and reactive power) control or P–F (active power and power factor) control, induction machine terminal voltage, and other variables that will need special handling to iteratively solve between machine equations and network equations.
2) Solve the network initial load flow equations for $\boldsymbol{y}_0$, i.e.,

$$0 = G(t_0 = 0, \boldsymbol{x}_0, \boldsymbol{y}_0) \tag{5.43}$$

3) Solve the next time step state variables $\boldsymbol{x}_1$ from state space equations,

$$\frac{\boldsymbol{x}_1 - \boldsymbol{x}_0}{\Delta t} = F(t_1 = t_0 + \Delta t, \boldsymbol{x}_1, \boldsymbol{y}_0) \tag{5.44}$$

4) With updated $\boldsymbol{x}_1$, repeat Step 2) to solve a new $\boldsymbol{y}_1$.
5) Continue to cycle the loop in Step 2) and Step 3) with a time step Δt incremented once at every new cycle until the specified simulation time t_{final} is reached,

$$0 = G(t_n, \boldsymbol{x}_n, \boldsymbol{y}_n)$$

$$\frac{\boldsymbol{x}_{n+1} - \boldsymbol{x}_n}{\Delta t} = F(t_{n+1}, \boldsymbol{x}_{n+1}, \boldsymbol{y}_n)$$

$$t_{n+1} = t_n + \Delta t$$

until,

$$t_n \geq t_{final} \tag{5.45}$$

Transient stability simulation involves solving a large number of mixed algebra and differential equations simultaneously. Fast, accurate, and robust numerical solution techniques are critical to obtain reliable results.

References

1 Williams, P.W. (1972). *Numerical Computation*. Thomas Nelson and Sons Ltd.

2 Tran, H.T.T., Nguyen, H.T., Vu, L.T., and Ojetola, S.T. (Feb 2024). Solving differential-algebraic equations in power systems dynamics with quantum computing. *IET Energy Conversion and Economics* 5 (1): 40–53.

3 Ascher, U.M. and Petzold, L.R. (1998). *Computer Methods for Ordinary Differential Equations and Differential-Algebraic Equations*. SIAM.

6

Examples of Practical I&C Power System Analysis

6.1 Objectives

Studies and analysis of industrial and commercial (I&C) power systems shall follow electrical system standards with clear objectives to validate operations and identify findings through well-designed scenario simulations. The analysis report shall recommend remedy or mitigation methods.

Electrical system standards applicable in the United States to I&C power system analysis include the following:

1) National Fire Protection Association (NFPA) standards, such as:
    a) NFPA 70 National Electrical Code
    b) NFPA 70E Standard for Electrical Safety in the Workplace
    c) etc.
2) Institute of Electrical and Electronics Engineers (IEEE) standards, such as:
    a) IEEE Std 399 IEEE Recommended Practice for Industrial and Commercial Power Systems Analysis (IEEE Brown Book)
    b) IEEE Std 3002.2 IEEE Recommended Practice for Conducting Load-Flow Studies and Analysis of Industrial and Commercial Power Systems
    c) IEEE Std 3002.3 IEEE Recommended Practice for Conducting Short-Circuit Studies and Analysis of Industrial and Commercial Power Systems
    d) IEEE Std 3002.7 IEEE Recommended Practice for Conducting Motor-Starting Studies and Analysis of Industrial and Commercial Power Systems
    e) IEEE Std 3002.8 IEEE Recommended Practice for Conducting Harmonic Studies and Analysis of Industrial and Commercial Power Systems
    f) etc.

Industrial and Commercial Power System Analysis Fundamentals and Practice, First Edition. J. J. Dai.

3) International Electrotechnical Commission (IEC) standards if the facility is within IEC areas, such as:
   a) IEC 60909 Short-circuit currents in three-phase a.c. systems
   b) IEC 61363-1 Electrical installations of ships and mobile and fixed offshore units – Part 1: Procedures for calculating short-circuit currents in three-phase a.c.
   c) etc.
4) Local standards and code of practice, if there are any.
5) Corporation specifications, if there are any.

This chapter demonstrates some common system analysis projects and examples in typical I&C power systems using commercial power system analysis software packages. They are:

1) Voltage drops study by applying load flow solution and simulation to show analysis findings and methods to improve system voltage profiles.
2) Power factor study also by applying load flow solution and analysis to show findings and methods to correct system power factors.
3) Over-current study by applying short-circuit solution and simulation to show analysis findings and adjustments to size circuit breakers.
4) A reactive power inrush study by applying motor starting solution and simulation to show analysis findings and mitigation to reduce motor starting inrush current and reactive power, as well as to meet system voltage regulation requirements.
5) Harmonic distortion study by applying harmonics solution and simulation to show analysis findings and mitigation to reduce harmonic distortions to meet distortion limitation requirements by standards.
6) Rotor angle swing study by applying transient stability solution and simulation to show analysis findings and protection to avoid instability issues.
7) Frequency decay study, which also by applying transient stability solution and simulation to show analysis findings and load-shedding control to maintain system frequency.

6.2 Voltage Drops Study

6.2.1 Introduction

6.2.1.1 Causes of Voltage Drops

Excessive voltage drops can interrupter system operations (for example, causing running motors to stall, preventing motors to successfully start, dropping of contactors) and/or affect equipment performance (for example, leading to voltage-sensitive devices, controls, and instrument malfunctions). Voltage drops in I&C power systems can be due to various reasons, for instance, poor system design can result in excessive voltage drops in the distribution network; thus, loads will not have sufficient voltage magnitude to maintain normal operations; motor starting can draw large reactive power and inrush current to add voltage drops in the feeder and transformer lasting for a few seconds; loss of power generation or increase of loads can shift the system operating condition away from the normal case and violate voltage constraints; lack of effective voltage regulations such as transformer taps, capacitor or reactive var compensator support, and generator automatic voltage regulator (AVR) response is also responsible for an under voltage profile; faults can instantaneously reduce the voltage in the system with different levels of severity.

6.2.1.2 Types of Voltage Drops

Based on the nature of system reactions, voltage drops are classified into steady state existence and transient state caused. The steady-state voltage drops exist persistently in the system even if there are no system transients. They are generally due to high impedance in the power distribution network and lack of voltage regulations. Resizing transformers, lines, and cables, and adding transformer taps, capacitor banks and AVR equipment at selected locations, reducing reactor impedance if there is any can improve voltage levels at designated buses and load terminals. The transient state voltage drops are mostly caused by time events such as starting large motors, impact adding a significant amount of load, losing on-site generation, disconnection from the grid, excitation and AVR system failures or poor performance, and different types of faults.

6.2.1.3 Voltage Drop Limitations

Excessive voltage drops can interrupt system operations (for normal operation, the common practice is to limit the permitted voltage variations on the highest voltage bus within plus or minus 1% and on the lower voltage buses within plus or minus 5%). Standards and corporate instructions specify voltage drop tolerances for different types of equipment at different values in more detail. At the steady state, in general, voltage drop from the source (incoming switchgear connecting to the utility) shall be limited to 5%. Table 6.1 shows a sample steady state allowable voltage drops at various feeders.

IEEE Standard 399-1997 lists the minimum allowable voltage (normal voltage less the voltage drops) at various locations during the motor starting transient state, which is shown in Table 6.2.

Table 6.1 Sample allowable voltage drops at various feeders.

Component	Voltage drops (%)
Main feeder	2
Motor feeder	3
Transformer feeder	3
Panel board feeder	1
Lighting branch and receptacle feeder	2

Table 6.2 Summary of critical system voltage levels when starting motors.

Locations	Minimum allowable voltage (% rated)
Terminals of starting motor	80%
Terminals of other motor that must accelerate	71%
AC contactor pick-up	85%
DC contactor pick-up	80%
Contractor hold-in	60–70%
Solid-static control devices	90%
Noticeable light flicker	3% change

6.2.1.4 Remedy for Voltage Drops

Load flow analysis is a tool used for addressing and analyzing voltage drop issues. First, load flow reports shall highlight and summarize voltage drop violations, their locations, and corresponding operation conditions. Common voltage drop issues are resolved throughout adjustable controls, including generator AVR, transformer tap changer, capacitor bank, and VAR compensators. Next, by configuring these controls individually or in combination, a satisfactory voltage profile at normal configuration in the load flow solution is achieved. Furthermore, the solution shall be extended to contingency configurations. This may require some readjustments and iterations via load flow simulation and analysis to control variables. Finally, other system and equipment operation constraints such as power factor, generator reactive power output limits, and branch equipment loading shall also be checked by load flow analysis to ensure they are all within limits.

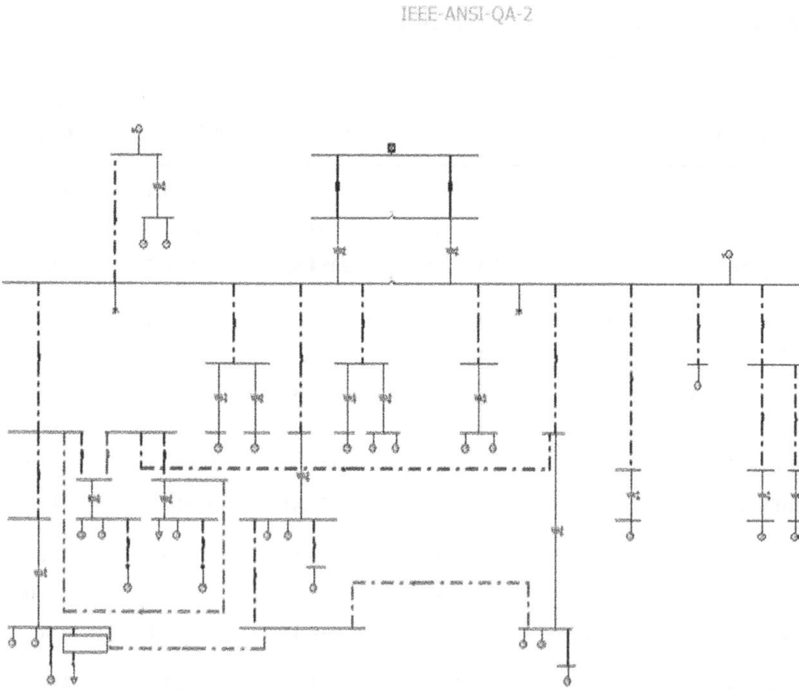

Figure 6.1 Single-line diagram for voltage drops analysis sample case.

6.2.2 A Voltage Drops Analysis Case

6.2.2.1 System Description

In this sample case, an industrial power system is illustrated for load flow analysis during a detailed study phase. Low-voltage conditions are initially identified, and a remedy action is taken to improve the voltage profile. The system single-line diagram is shown in Figure 6.1, and the system power supply and load data are shown in Table 6.3.

6.2.2.2 Initial Simulation Results

The initial load flow analysis by computer software identifies some low bus voltage conditions, as listed in Table 6.4. These low voltage buses as well as connected equipment are displayed in red color [refer online] for better visualization of the single-line diagram, as shown in Figure 6.2.

A detailed view of one of the low-voltage buses with ID 18:T6-SEC is shown in Figure 6.3 indicating the bus voltage magnitude is 0.948 pu or 94.8%, below the required 95%.

Table 6.3 System data summary for voltage drops analysis sample case.

Total summary	kW	kvar	kVA	Power factor (PF) (%)
Sources (swing)	4135.59	3186.16	5220.61	79.22
Generators	25312.50	8898.80	26831.16	94.34
Total generation	**29448.09**	**12084.95**	**31831.37**	**92.51**
Load read (non-adjusted)	2400.00	1200.00	2683.28	89.44
Load used (adjusted)	2400.00	1200.00	2683.28	89.44
Shunt capacitors (adjusted)	0.00	−3943.87	3943.87	0.00
Shunt reactors (adjusted)	0.00	0.00	0.00	0.00
Motors	26776.14	12968.28	29751.26	90.00
Total loads	**29176.14**	**10224.41**	**30915.78**	**94.37**
Cable capacitance	0.00	−0.13	0.13	0.00
Line capacitance	0.00	0.00	0.00	0.00
Total shunt capacitance	**0.00**	**−0.13**	**0.13**	**0.00**
Line losses	5.21	11.11	12.27	42.45
Cable losses	57.22	76.64	95.64	59.82
Transformer load losses	209.51	1772.91	1785.25	11.74
Transformer no-load losses	0.00	0.00	0.00	0.00
Total losses	**271.94**	**1860.67**	**1880.43**	**14.46**

Table 6.4 Low-voltage bus tabulation.

Node Id	Base Voltage (kVLL)	V (kVLL)	V (p.u.)	Angle V (°)
18:T6-SEC	0.480	0.4	0.933	54.04
23:T6MCC	0.480	0.4	0.933	54.03
21:T9-SEC	0.480	0.5	0.946	55.20
36:T14-SEC	0.480	0.5	0.946	55.13
28:T10-SEC	0.480	0.4	0.932	-4.18
33:T10MCC	0.480	0.4	0.930	-4.19
41:LGTS	0.480	0.4	0.926	-4.21

Load Flow - Buses outside limits X | Load Flow - Buses | Load Flow - Gene

	Node Id	Base Voltage (kVLL)	V (kVLL)	V (p.u.)	Angle V (°)
1	18:T6-SEC	0.480	0.4	0.933	54.04
2	23:T6MCC	0.480	0.4	0.933	54.03
3	21:T9-SEC	0.480	0.5	0.946	55.20
4	36:T14-SEC	0.480	0.5	0.946	55.13
5	28:T10-SEC	0.480	0.4	0.932	-4.18
6	33:T10MCC	0.480	0.4	0.930	-4.19

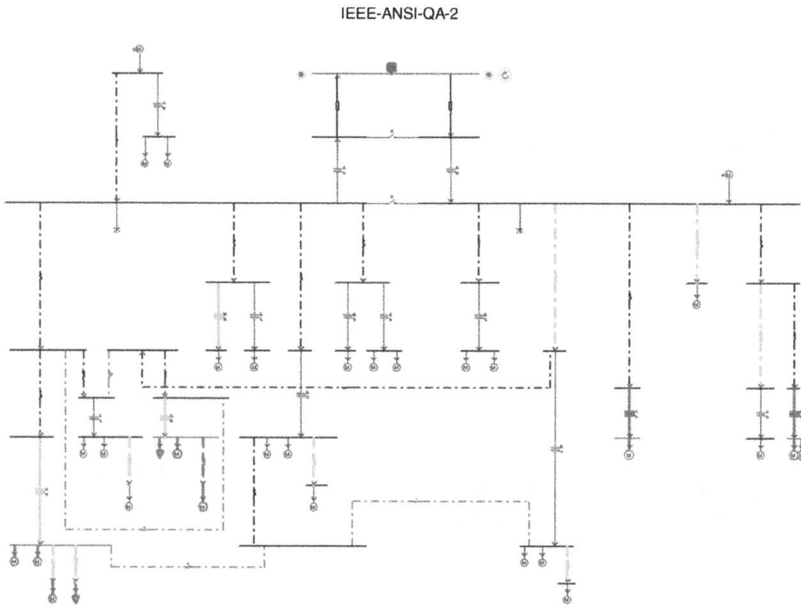

Figure 6.2 Load flow results for low-voltage buses and connected equipment.

6.2.2.3 Remedy Action

Several remedy actions can be taken to relieve the low voltage conditions, such as through generator AVR regulation, transformer tap adjustment, capacitor bank and static var compensator installation, distribution network impedance reduction, etc. In this example, an upstream transformer tap changer can be an effective solution. The upstream 2 MVA transformer originally had no tap setting. Adjust the secondary side tap (connecting to bus 18:T6-SEC) from 100% to 102%, as shown in Figure 6.4 with an intention to raise bus 18:T6-SEC by around 2% and rerun the load flow analysis. Figure 6.5 shows the improved bus 18:T6-SEC voltage to 96.7% above the limit.

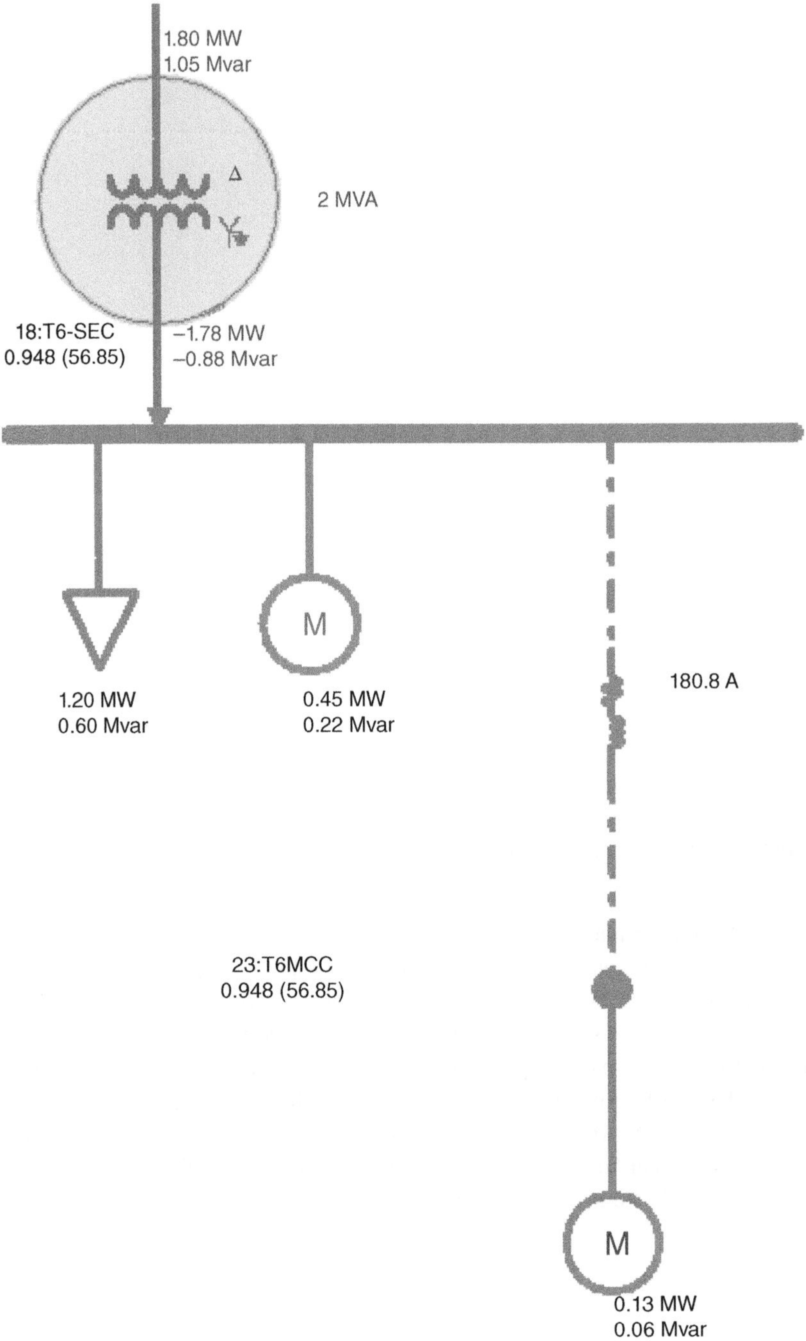

Figure 6.3 Detailed view of a low-voltage bus.

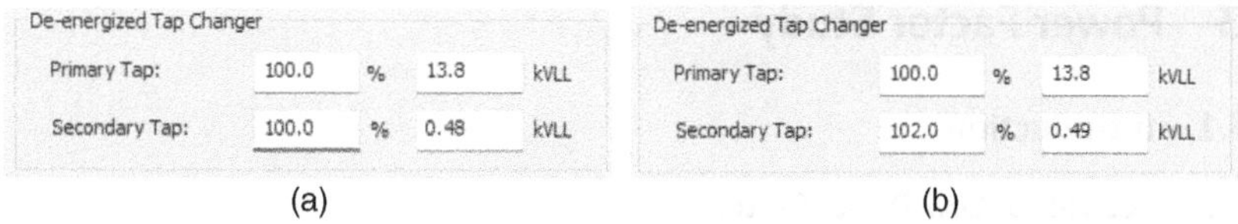

Figure 6.4 Adjustment of the upstream transformer tap setting on the secondary: (a) before adjustment and (b) after adjustment.

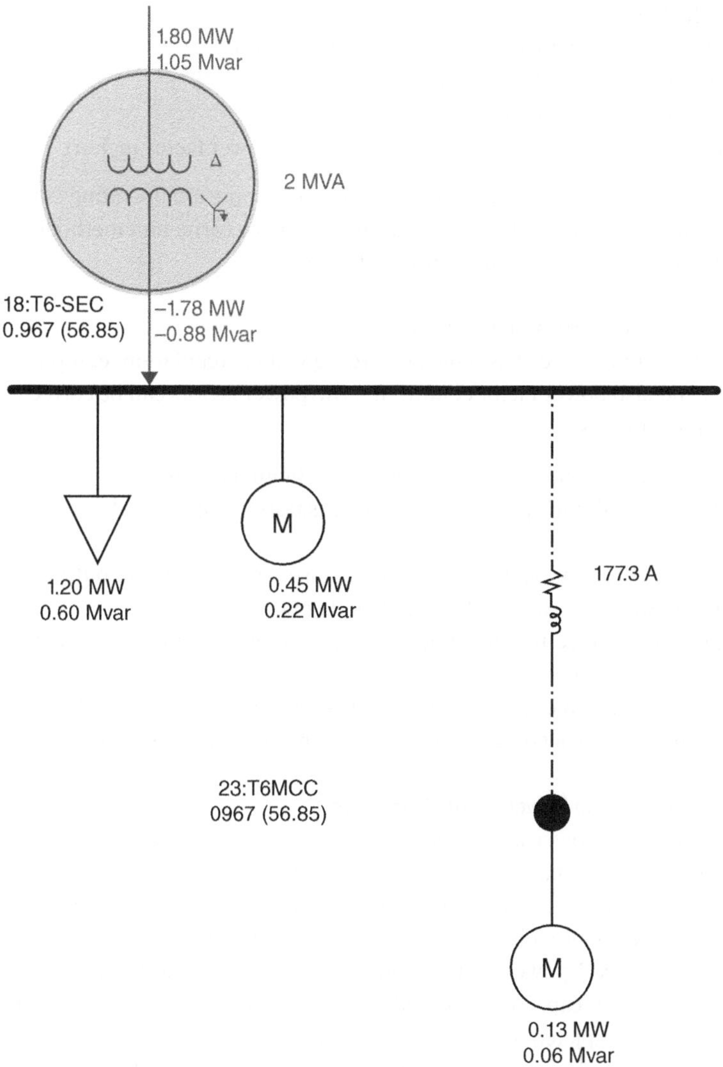

Figure 6.5 Improved bus 18:T6-SEC voltage.

6.3 Power Factor Study

6.3.1 Introduction

6.3.1.1 Causes of Low-Power Factor

Low power factor conditions existing in I&C power systems could be due to one or combined situations in the system:

1) Heavy inductive loads.
2) High transformer reactance over resistance (X/R) ratio.
3) Improper size of current limiting reactors.
4) Lack of reactive power supply.
5) Light motor loading (motor operating in the lower power factor region).

Some of the above causes are design originated, and some are operating condition related, and some are due to both. Thus, power factor correction methods are developed to tackle these root causes from both bases.

6.3.1.2 Power Factor Correction Methods

System load flow analysis reveals issues regarding voltage regulation, equipment overloading, and also power factor condition. Low-power factors can be corrected using several techniques:

1) Installing capacitors and enabling automatic switching control.
2) Installing and configuring static var compensators to operate at capacitive range.
3) Increasing synchronous generators and condensers reactive power output.
4) Replacing induction motors with synchronous motors.
5) Leveraging capacitance in harmonic filters to compensate inductive loads at the fundamental frequency.
6) Programming adjustable speed drive to reduce motor reactive power demand.
7) Reducing inductive reactance of transformers and reactors if possible.

6.3.1.3 Special Notes on Power Factor Correction

Note that some of the power factor correction approaches require system design modifications and may create new issues such as harmonic parallel resonance. Therefore, repeated load flow and other analyses shall be engaged to iteratively resolve the issues while providing the final satisfactory design and solution. Some other notes for the I&C power systems on this subject include if applying the power factor correction capacitors with adjustable speed drive, careful harmonic analysis shall be completed to fully cover harmonic injection spectrum and loading demand; the power factor correction assemblies shall be able to operate continuously at full rated voltage and current, and provide effective power factor corrections under different operating scenarios, preferably a minimum of 95% lagging power factor throughout the network and close to unity at the utility interfacing point; switchable power correction assemblies are more flexible for systems that frequently varying loading conditions. Special protections are also needed to protect the capacitors during energizing, switching, disconnecting, and system fault conditions.

6.3.2 A Power Factor Correction Analysis Case

6.3.2.1 System Description and Initial Simulation

Continue with the previous sample case and check point of common coupling (PCC) (utility source or swing bus) power factor to find out it is only 79.22% as shown in Table 6.5. Most utility request power factors at PCC being 98% or above. Failure to meet the requirement will result in penalties from the utility due to lower

Table 6.5 Load flow summary with interconnection power factor.

Total summary	kW	kvar	kVA	PF (%)
Sources (swing)	4135.63	3186.23	5220.67	79.22
Generators	25312.50	8898.71	26831.13	94.34
Total generation	**29448.13**	**12084.94**	**31831.40**	**92.51**
Load read (non-adjusted)	2400.00	1200.00	2683.28	89.44
Load used (adjusted)	2400.00	1200.00	2683.28	89.44
Shunt capacitors (adjusted)	0.00	−3943.87	3943.87	0.00
Shunt reactors (adjusted)	0.00	0.00	0.00	0.00
Motors	26776.14	12968.28	29751.26	90.00
Total loads	**29176.14**	**10224.41**	**30915.78**	**94.37**
Cable capacitance	0.00	−0.13	0.13	0.00
Line capacitance	0.00	0.00	0.00	0.00
Total shunt capacitance	**0.00**	**−0,13**	**0.13**	**0.00**
Line losses	5.24	11.17	12.33	42.45
Cable losses	57.18	76.58	95.57	59.82
Transformer load losses	209.57	1772.90	1785.24	11.74
Transformer no-load losses	0.00	0.00	0.00	0.00
Total losses	**271.98**	**1860.65**	**1880.42**	**14.46**

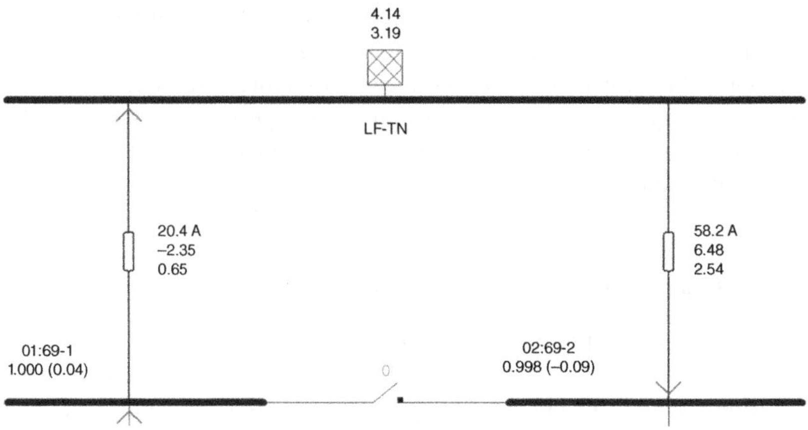

Figure 6.6 Utility active and reactive power supply.

power factor resulting in higher reactive power demand to reduce the operation profit.

Examination of the load flow analysis displays on single-line diagram in Figure 6.6 shows following power supply from the utility:

$$\text{Active power MW} = 6.48 - 2.35 = 4.13$$

$$\text{Reactive power Mvar} = 2.54 + 0.65 = 3.19$$

from which the power factor can be derived:

$$\text{Power factor} = \frac{4.13}{\sqrt{4.13^2 + 3.19^2}} = 0.79 \text{ or } 79\% \text{ lagging}$$

6.3.2.2 Remedy Action

To remedy the low-lagging power factor issue, either to increase the generator reactive power output, or add capacitor bank or var compensator. Further detailed check of the two on-site generator load flow analysis results confirms both generators are already at their maximum reactive power output, thus increase capacitor size seems the reasonable solution. The primary design of system proposes to have a capacitor bank ID CAP1 connected to 13.8 kV bus as shown in Figure 6.7.

This capacitor is initially rated at 3 × 670 kvar as suggested value and now try to double its size to 3 × 1340 kvar so more reactive power can be obtained from inside the system instead of importing from the utility, as shown in Figure 6.8.

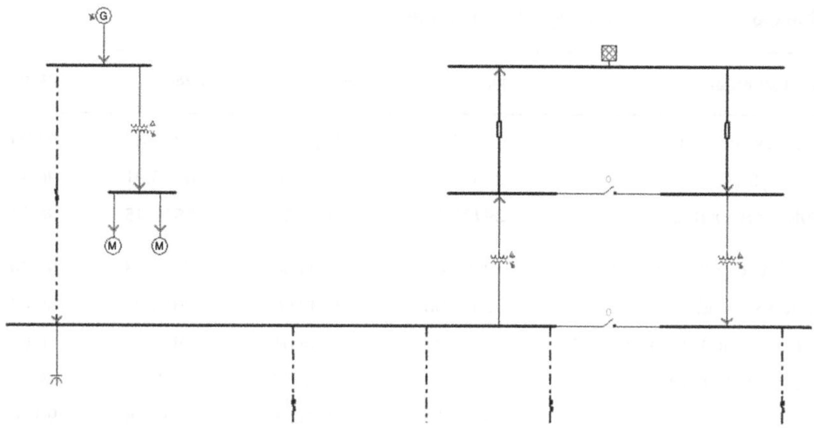

Figure 6.7 Capacitor location.

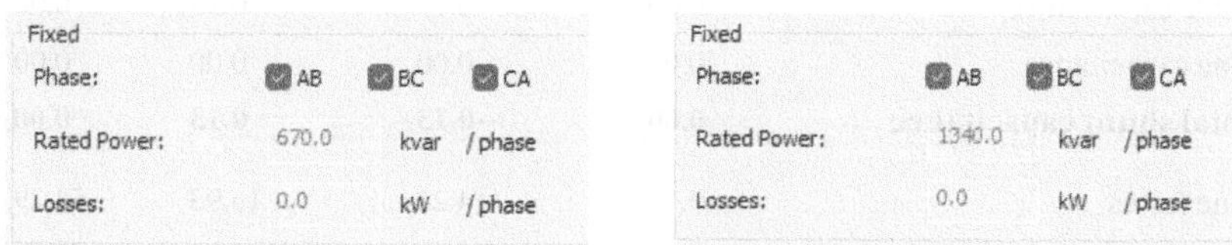

(a) (b)

Figure 6.8 Capacitor size: (a) initial rating, (b) increased rating.

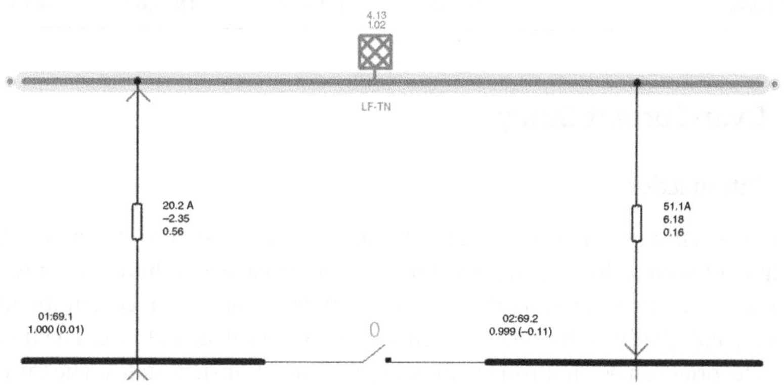

Figure 6.9 Utility active and reactive power supply after the remedy.

6.3.2.3 Validation

Rerun of load flow analysis with the increased capacitor size verifies the utility power factor is successfully improved to 97%., It is an 18% increase. Table 6.6 and Figure 6.9 are from the latest load flow analysis with the remedy.

Table 6.6 Load flow summary after the remedy.

Total summary	kW	kvar	kVA	PF (%)
Sources (swing)	4133.79	1020.25	4257.83	97.09
Generators	25312.50	6967.44	26253.91	96.41
Total generation	**29446.29**	**7987.69**	**30510.45**	**96.51**
Load read (non-adjusted)	2400.00	1200.00	2683.28	89.44
Load used (adjusted)	2400.00	1200.00	2683.28	89.44
Shunt capacitors (adjusted)	0.00	−7985.01	7985.01	0.00
Shunt reactors (adjusted)	0.00	0.00	0.00	0.00
Motors	26776.14	12968.28	29751.26	90.00
Total loads	**29176.14**	**6183.26**	**29824.15**	**97.83**
Cable capacitance	0.00	−0.13	0.13	0.00
Line capacitance	0.00	0.00	0.00	0.00
Total shunt capacitance	**0.00**	**−0.13**	**0.13**	**0.00**
Line losses	9.19	14.22	16.93	54.29
Cable losses	49.77	70.08	85.95	57.90
Transformer load losses	211.18	1720.26	1733.18	12.18
Transformer no-load losses	0.00	0.00	0.00	0.00
Total losses	**270.15**	**1804.56**	**1824.67**	**14.81**

6.4 Over-Current Study

6.4.1 Introduction

Types of over-current issues. Over-current issues can be related to: (1) network equipment such as line, cable, transformer, reactor, capacitor, harmonic filter, (2) load equipment such as motor and other types of power consumption devices, and (3) protective devices such as a breaker, switch, and fuse. The first two categories can occur during both steady state and transient states. The third category concerns about the device withstand and interrupting capabilities under short-circuit fault conditions, which is discussed here. Protective device over-current condition is also referred to as overstressed protective device.

Over-current to circuit breakers. The acceptable fault current duty levels for breakers shall be limited under the short-circuit ratings of the switchgear at all stages of the fault, namely, momentary, interrupting, and steady state (continuous) stages. Due to the dc component, fault currents are asymmetrical; thus, properly overseeing of the calculated fault currents and comparing them with the switchgear nameplate specified breaker ratings is vital. All connected operating equipment in the network shall be considered in calculations to reflect the maximum possible fault currents. Three-phase fault and single-phase-to-ground fault shall always be simulated to calculate fault current at three stages. Other types of faults shall also be evaluated. The industry practices identifying the fault current levels are the calculated fault current is 90% above the protective device ratings. Above this level, situations need to be carefully assessed. Short-circuit analysis shall identify and summarize all over-current breakers and percentages of over-current versus the ratings at different fault duty stages.

Remedy of circuit breaker over-current. Over-current remedies for circuit breakers include two options, the first one is upsizing the breakers for higher rating to cover the fault current duties; the second is increasing reactance of the network component along the fualt current paths to reduce the fault current, typically by adding a current-limiting reactor, or chose higher reactance for an existing current-limiting reactor. Sometimes, the transformer connected upstream to the over-current breaker can be specified with higher winding impedance to limit the fault current feeding to the breaker.

6.4.2 An Over-Current Analysis Case

6.4.2.1 System Description and Fault Currents

Short-circuit analysis is conducted for a sample industrial system illustrated in Figure 6.10.

The short-circuit calculation is per ANSI standard. Bus fault results including closing and latching (at ½ cycle), peak, and contact parting symmetrical current for three-phase and line-to-ground faults to all buses individually are displayed in Figure 6.11.

6.4.2.2 Over-Current Breaker Sizing

Detailed fault information display for Bus B6 at 13.8 kV is found in Figure 6.12, from which it is identified that for a three-phase fault (LLL), the closing and latching duty is 83.9 kA, and the contact parting duty is 46.3 kA.

Based on the bus fault duties, a circuit breaker to isolate the fault from Bus B6 is to be selected from the inventory. Assume the first attempt is to choose a breaker model 15PV50STD 1200 as shown in Figure 6.13a. This breaker has maximum symmetrical interrupting capability of 50.0 kA which will meet the contact parting duty of 46.3 kA. However, its closing and latching capacity is 80.0 kA which will not cover 83.9 kA closing and latching fault duty. A new selection is needed.

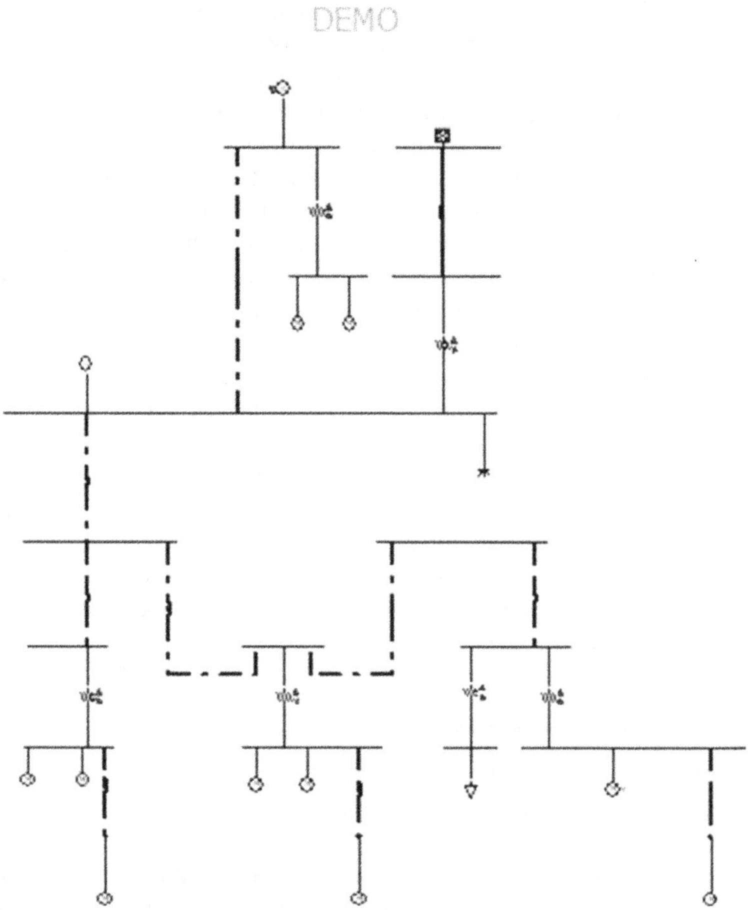

Figure 6.10 Short-circuit over current analysis sample system.

Reports

Search

	Node Id	Prefault kV (kV)	LLL ½ cycle (kA)	LLL Peak (kA)	LLL CP Sym (kA)	LG ½ cycle (kA)	LG Peak (kA)	LG CP Sym (kA)
1	100:UTIL-69	69.000	8.442593	14.247439	5.485314	5.861915	9.883645	3.831926
2	B3	69.000	7.379708	12.573904	4.992333	4.496390	7.731259	3.227065
3	B4	13.800	174.299435	303.322306	58.207207	36.535215	61.596240	20.802960
4	B1	13.800	83.433460	145.279371	42.856076	25.634347	43.550853	15.703571
5	B5	13.800	92.777651	156.264247	47.661447	28.320953	49.144022	18.998651
6	B7	13.800	42.162229	65.747757	30.370605	18.090393	31.225517	14.931440
7	B8	13.800	87.945373	147.164996	46.536086	27.557842	47.908992	18.783152
8	B6	13.800	87.354652	146.074367	46.388553	27.464001	47.755381	18.754064
9	B9	13.800	80.428312	135.199078	44.113351	26.748435	46.527485	18.372013

ANSI Short-Circuit - High Voltage Buses Current Summary | ANSI Short-Circuit - Low Voltage Buses Current Summary

Figure 6.11 Short-circuit calculation display.

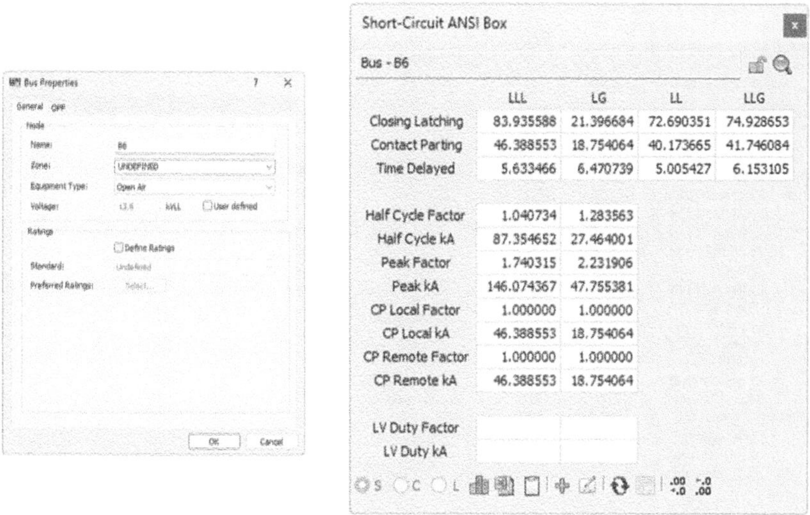

(a) (b)

Figure 6.12 Fault duties for bus B6. (a) Bus information. (b) Bus fault duties.

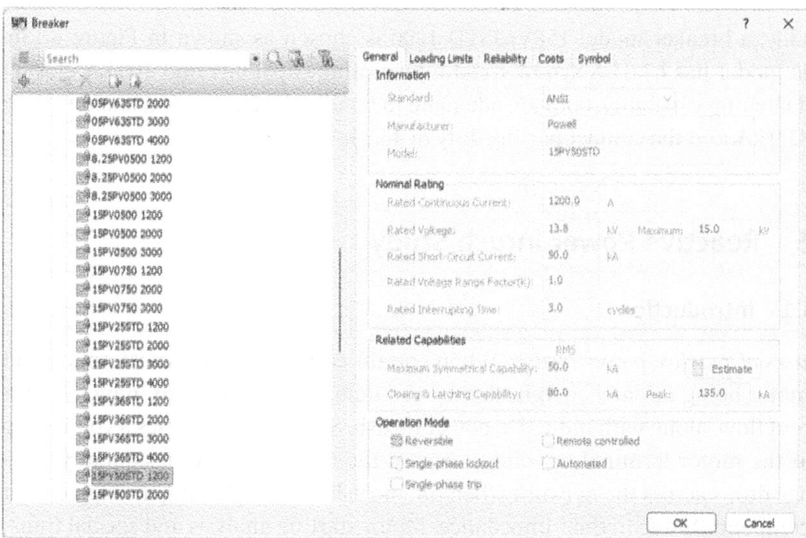

(a)

Figure 6.13 Breaker sizing. (a) Undersized breaker and data. (b) Correct breaker size and data.

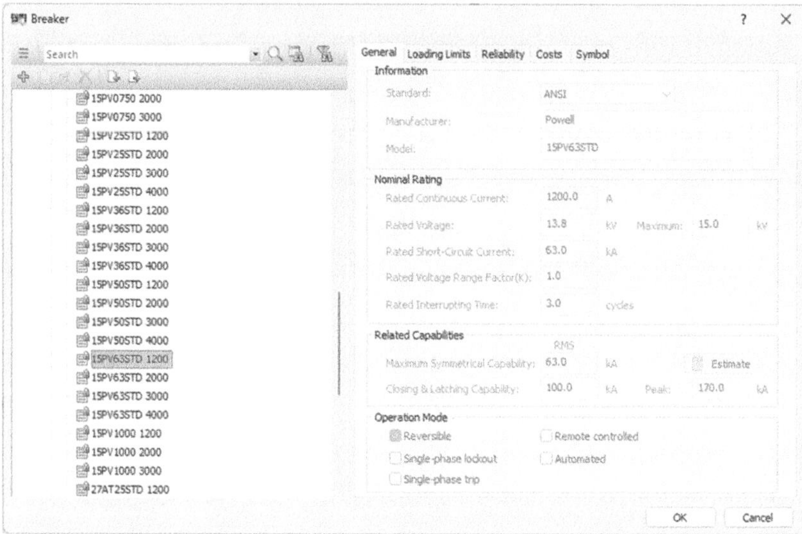

(b)

Figure 6.13 (*Continued*)

Finally, a breaker model 15PV63STD 1200 is chosen as shown in Figure 6.13b. This model has 63.0 kA symmetrical interrupting capability and 100 kA closing and latching capability, both are adequate to handle the closing and latching duty of 83.9 kA and the contact parting duty of 46.3 kA.

6.5 Reactive Power Inrush Study

6.5.1 Introduction

Causes of reactive power inrush. When certain equipment like transformers and motor being initially energized, they can demand large in-rush inductive current flow, along with inductive power inrush. Such current causes a voltage dip at the motor terminal or voltage drop at the transformer. Consequently, voltage drops across the network when the in-rush current flowing through circuit components due to their impedance. Motor starting analysis and special transformer energization analysis shall study and report the magnitude of reactive power and current inrush and identify associated voltage drops and equipment overloading issues.

Remedy of reactive power inrush. In the motor starting case, the reactive power inrush-induced issues can be mitigated by two approaches. The first approach is to reduce the motor starting voltage. Since the starting motor draws an inrush current directly proportional to its terminal voltage, therefore, at a lower voltage, the motor will require less current, thereby the voltage dip will be reduced. Autotransformer starters are very effective in reducing starting voltage. The standard voltage tap setting for autotransformer starters ranges from 50% to 80% of the normal rated voltage. The motor starting analysis shall select the proper voltage tap to lower current inrush, and meet voltage regulation requirements. The second approach to reducing high inrush reactive power is to use a capacitor in parallel with the motor during the start. The motor inrush inductive power will be offset by the capacitive power provided by the capacitor thus the net reactive inrush power will be reduced, and the motor can still have the near full voltage to start. Motor starting analysis shall recommend the optimal solutions with parameters of the starting devices and full performance validation.

Further remedy to reactive power inrush. Since the starting motor reactive power inrush can affect the system at various locations to experience voltage difficulties, if it is possible, leveraging off-nominal tap settings for distribution transformers in the network can provide an effective and economical solution for voltage dips. Motor starting analysis shall monitor bus voltages across the network, adjust and validate distribution transformer taps, and recommend settings. This process may need load flow analysis in combination to ensure both low-voltage issue during the reactive power inrush and high-voltage issue after motors are started due to off-nominal tap settings are successfully resolved.

6.5.2 A Reactive Power Inrush Analysis Case

6.5.2.1 System Description

In this sample motor starting analysis case, a 450 HP 480 V three-phase motor PM-101 is being started. Attention is given to the motor starting current, reactive power, speed, terminal voltage and voltages at two adjacent buses, one is bus ID B15 at the motor terminal, and the other is a load bus ID B12 as shown in Figure 6.14.

The starting motor is modeled dynamically in a single-cage equivalent circuit. Its nameplate and circuit model data are found from Figure 6.15a and b.

6.5.2.2 Comparison of Starting Methods

Two starting methods are compared. The first is the direct cross-line start, and the second is a soft start. Starting configurations of those two starting methods are presented in Figure 6.16a and b, respectively. The soft starting method controls motor voltage over the time during the starting to achieve desirable inrush current and reactive power reduction. In this sample case, it applies a ramp up voltage with a controllable kick start voltage to the motor as illustrated in Figure 6.16b.

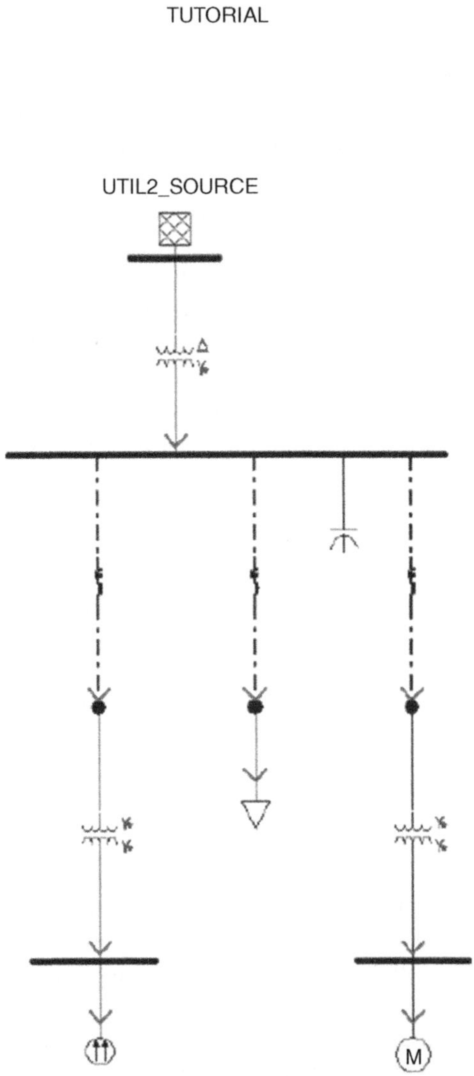

Figure 6.14 Motor start analysis sample case.

Simulation results from the two starting methods are plotted in Figures 6.17 and 6.18. Clearly, the motor starting inrush current and reactive power are both very high when applying the direct cross-line start method in Figure 6.17a. As a result, motor terminal voltage curve V (PM-101) which looks like the inverse of starting current in Figure 6.17b suffers a prolonged voltage dip. Motor bus voltage V (B-15) essentially is the same as motor terminal voltage V (PM-101) as shown in Figure 6.17b and c.

In comparison, the soft start applies a ramp voltage to the motor terminal which forces motor starting current and reactive power demand to follow the voltage

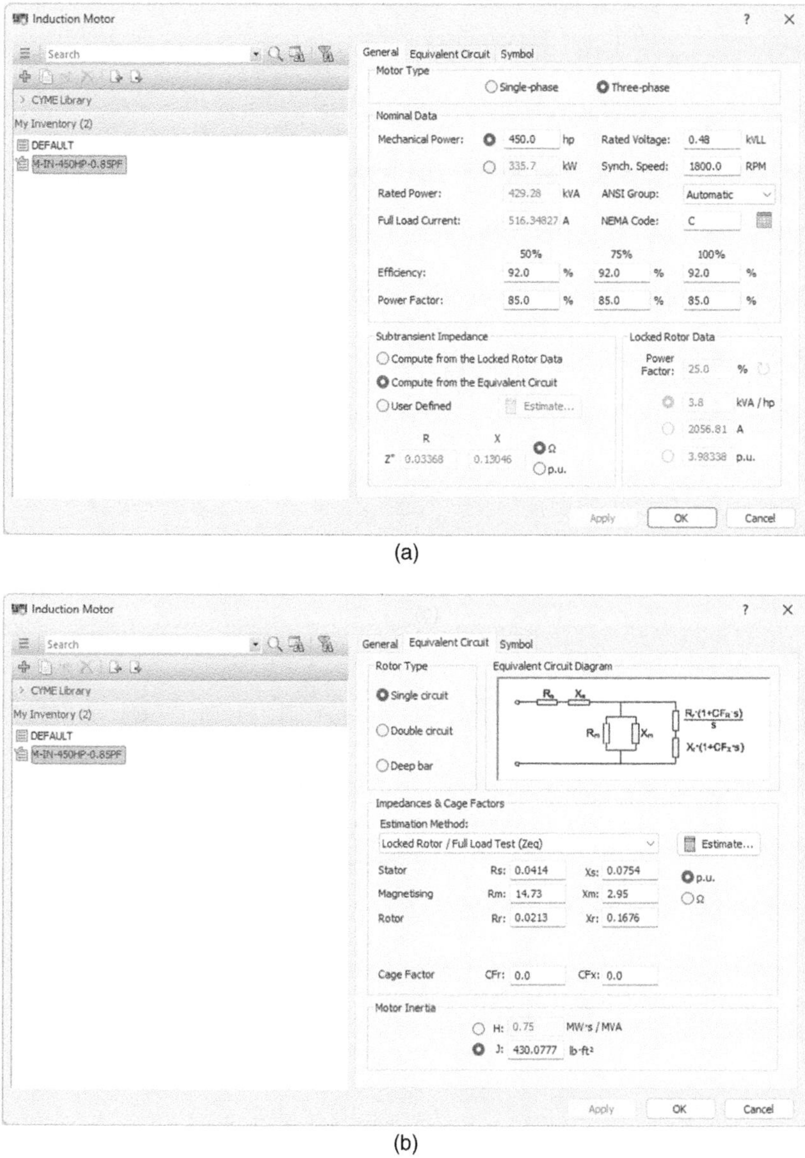

Figure 6.15 Starting motor data: (a) Nameplate. (b) Equivalent circuit model.

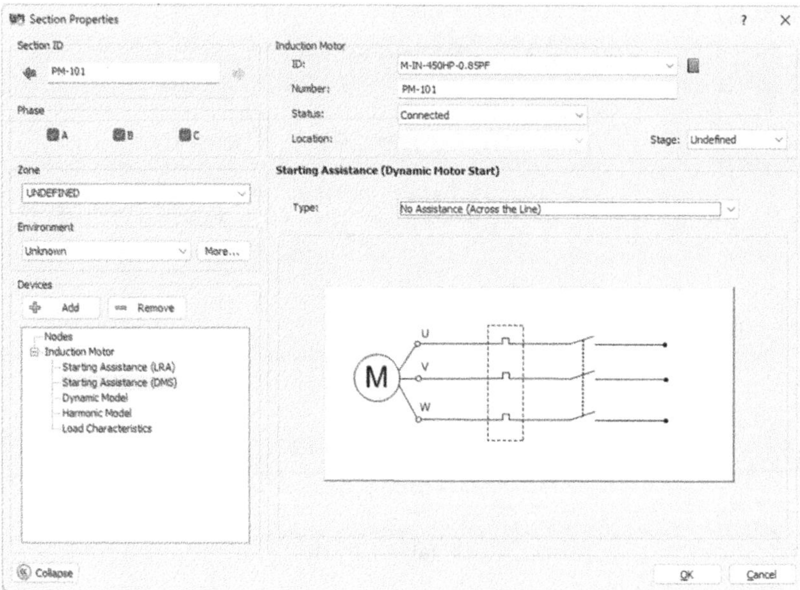

(a)

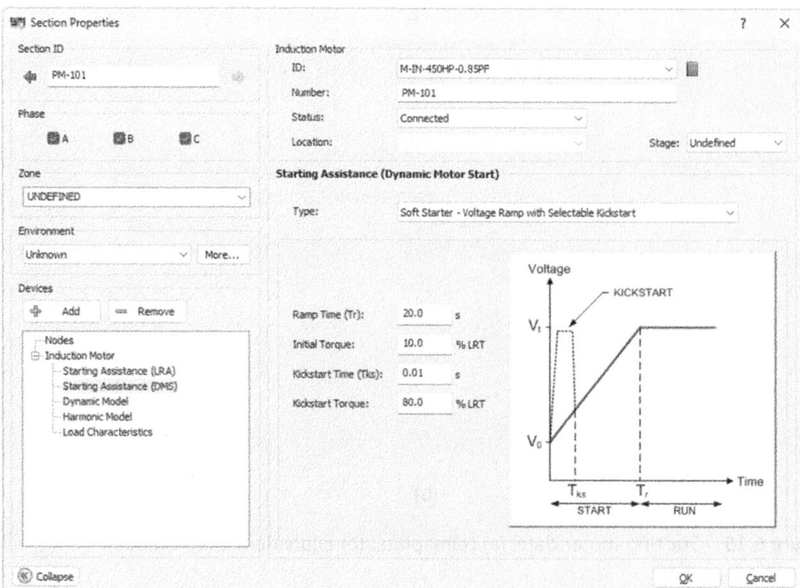

(b)

Figure 6.16 Motor starting configurations: (a) Cross-line start. (b) Soft start.

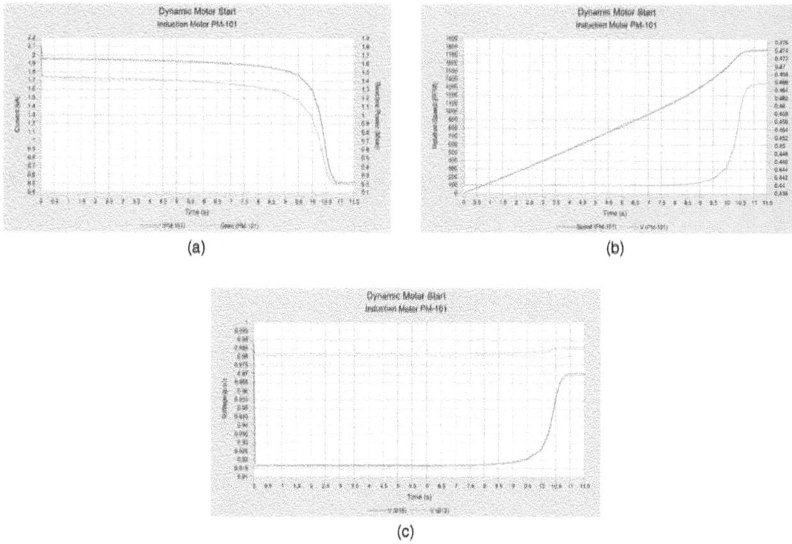

Figure 6.17 Motor starting responses from the cross-line start: (a) Motor current (blue [refer online]) and reactive power (green [refer online]). (b) Motor speed (blue [refer online]) and terminal voltage (green [refer online]). (c) Bus B15 (blue [refer online]) and B12 (green [refer online]) voltages.

curve, as demonstrated in Figure 6.18a thus the network avoids the initial voltage drop impact when the motor is started. The motor terminal voltage V (PM-101), which is also bus B15 voltage V (B-15), is controlled as shown in Figure 6.18b and c.

However, it needs to be pointed out that reducing starting voltage limits motor inrush current and reactive power which affects motor starting time. Check motor speed curves in plots (b) in Figures 6.17 and 6.18 to find out that the starting time for the specific soft start takes about 20 seconds, which is almost double the starting time by the direct cross-line start. This is the trade-off of less network voltage dip and sudden impact on other operating loads.

6.6 Harmonic Distortion Study

6.6.1 Introduction

6.6.1.1 Harmonic Distortion Limits

Harmonic distortion issues shall be investigated through harmonic analysis, starting in the preliminary design phase and completing in the detailed design phase. The analysis models all operating harmonic sources or nonlinear elements that distort voltage or current and thus effectively become harmonic sources. The analytic results shall provide bus voltage and branch current total and individual

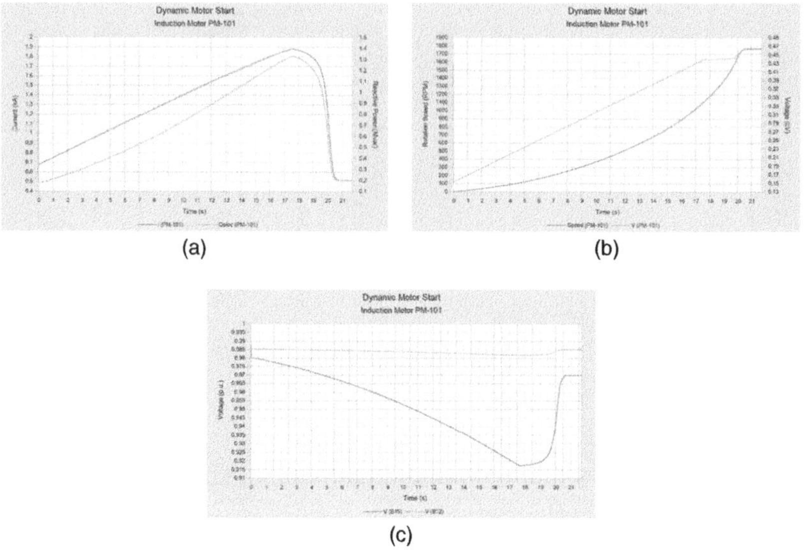

Figure 6.18 Motor starting responses from the soft start: (a) motor current (blue [refer online]) and reactive power (green [refer online]), (b) motor speed (blue [refer online]) and terminal voltage (green [refer online]), (c) bus B15 (blue [refer online]) and B12 (green [refer online]) voltages.

harmonic distortion indices, including those at PCC. PCC harmonic distortion limits are specified in relevant standards [1, 2], For example:

1) Total harmonic voltage distortion < 5%
2) Individual harmonic voltage distortion < 3%
3) Total and individual harmonic current distortion < 5% or less
4) Voltage notch depth < 20%
5) Other harmonic sensitive devices and instruments or those that require high-power quality may have additional requirements.

6.6.1.2 Mitigation of Harmonic Distortions

Harmonic distortion violations shall be mitigated. Standard methods include design improvements, turning and installation of filters, employment of high-pulse drive, specification and configuration of phase-shift transformers, sizing and installation of chocks, etc. The final harmonic analysis report shall include the mitigation equipment and the turned parameters, with validations to prove harmonic distortions under all loading conditions below the standard and corporat limits.

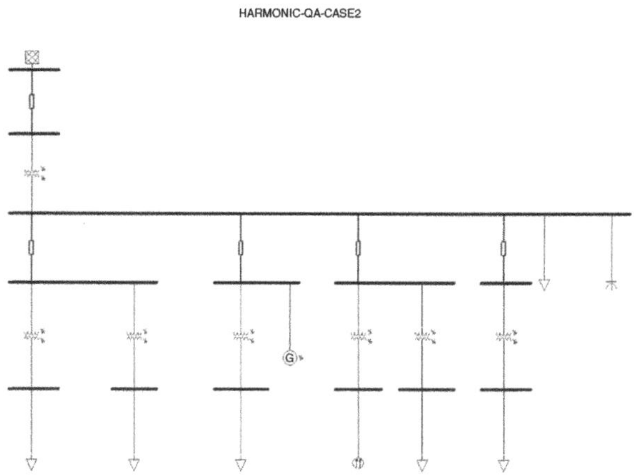

Figure 6.19 Harmonic analysis system single-line diagram.

6.6.2 A Harmonic Distortion Analysis Case

6.6.2.1 System Description

A sample industrial power system in 60 Hz is connected to a utility grid line at 69 kV level and has a number of loads at 2.4 kV buses after a couple of voltage transformations, as shown in Figure 6.19. A 3.2 MVA nonlinear load is a harmonic source that needs to be studied. Figure 6.20 provides harmonic information on this VFD load. It appears the harmonic is from a six-pulse switching power converter.

6.6.2.2 Initial Study Results

Run a harmonic load flow simulation and frequency scan to the system. Figure 6.21 shows harmonic penetration into the network with the red-colored [refer online] locations in high-harmonic distortions. For example, the secondary side of the distribution transformer bus ID SEC_13.8KV total voltage harmonic distortion (THD) exceeds 5%.

Figure 6.22 are the plots for bus ID SEC_13.8KV: (a) is bus voltage magnitude versus frequency which indicates relatively high-harmonic contents at the fifth, and seventh harmonic order; (b) is bus voltage waveforms for three phases which show noticeable distortion from the ideal sinusoidal shape; and (c) is from the frequency scan study which identifies one parallel resonance around 420 Hz or close to the seventh harmonic.

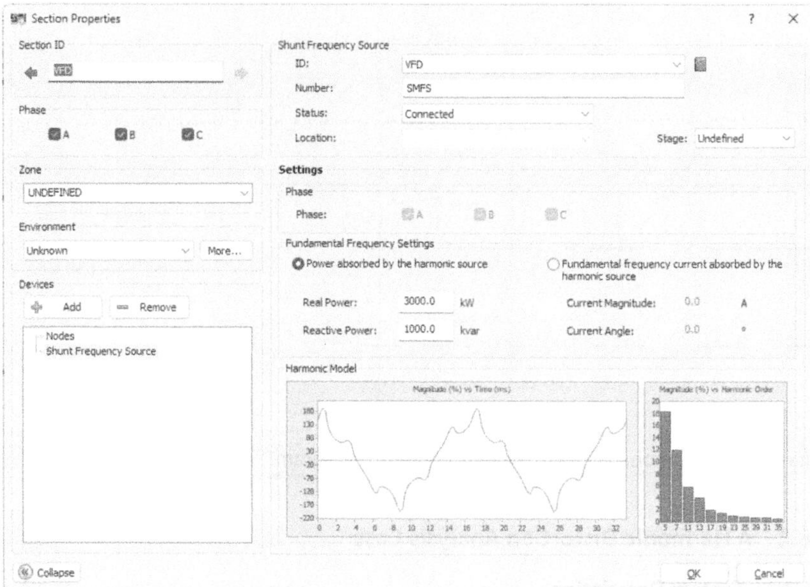

Figure 6.20 Harmonic current source waveform and spectrum.

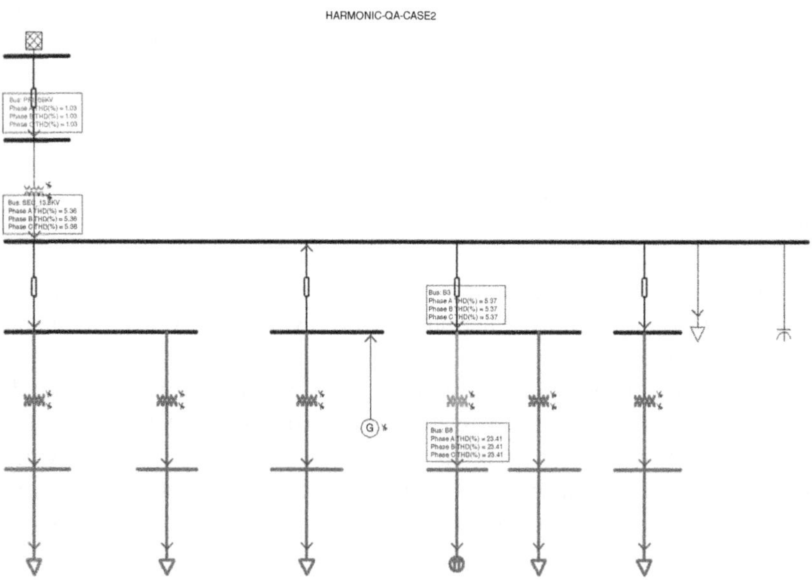

Figure 6.21 Harmonic analysis results display.

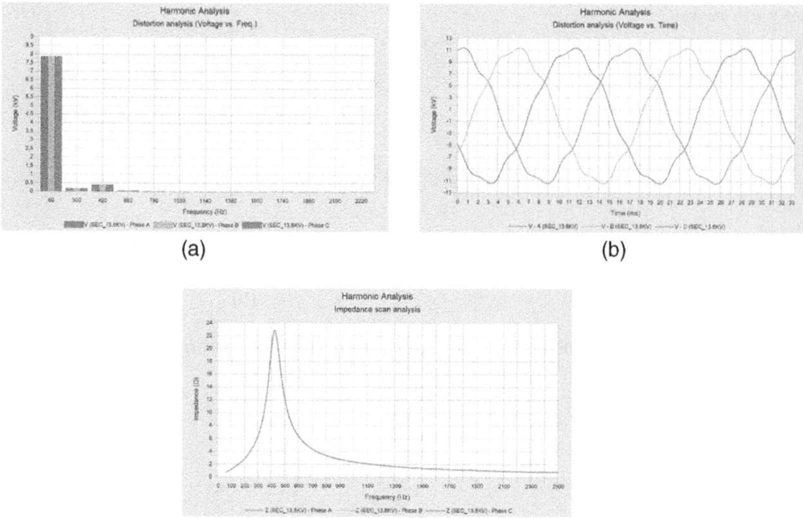

(a) (b) (c)

Figure 6.22 The plots of harmonic analysis results: (a) Voltage versus frequency at bus ID SEC_13.8KV. (b) Voltage waveform at bus ID SEC_13.8KV. (c) Impedance frequency scan at bus ID SEC_13.8KV.

6.6.2.3 Mitigation Actions

To mitigate harmonic distortions, passive filters are one of the typical solutions for I&C power systems. Since there is a fifth harmonic current component injecting into bus ID SEC_13.8KV from the downstream connection harmonic source, adding a single tuned filter to the fifth harmonic on this bus to shunt the fifth harmonic current (because the filter offers a low-impedance path to the current at the tuned frequency) is an option. Suppose it is estimated to add a total of 6000 kvar capacitance to the network from the power factor correction consideration point of view and chose a quality factor 50 for tuning point selectivity (higher quality factor offers sharper the selectivity). By assigning the tuning frequency slightly below the fifth harmonic, say at 4.7th harmonic due to the detuning effect from the inductive load in the system, and assuming the filter connected bus voltage is 13.8 kV, filter parameters R, L, and C are computed. The above data is shown in Figure 6.23a and b.

6.6.2.4 Mitigation Validation

After the designed filter is added to the bus ID SEC_13.8KV and rerun the harmonic load flow simulation and frequency scan, updated results are displayed in Figure 6.24. The results show voltage THD at the bus ID SEC_13.8KV is reduced

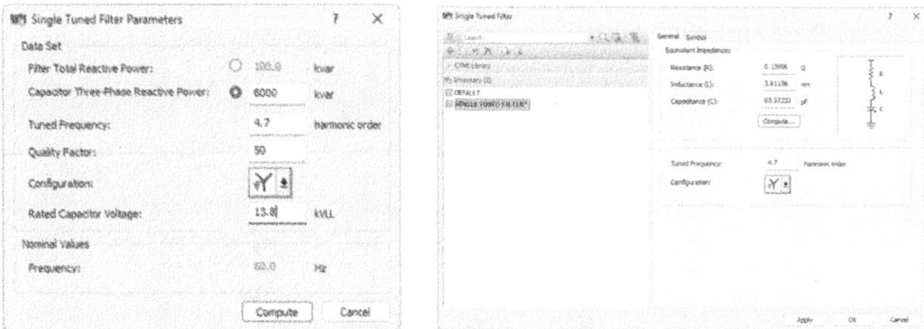

(a) (b)

Figure 6.23 Single-tuned filter data: (a) Filter rating. (b) Filter parameters.

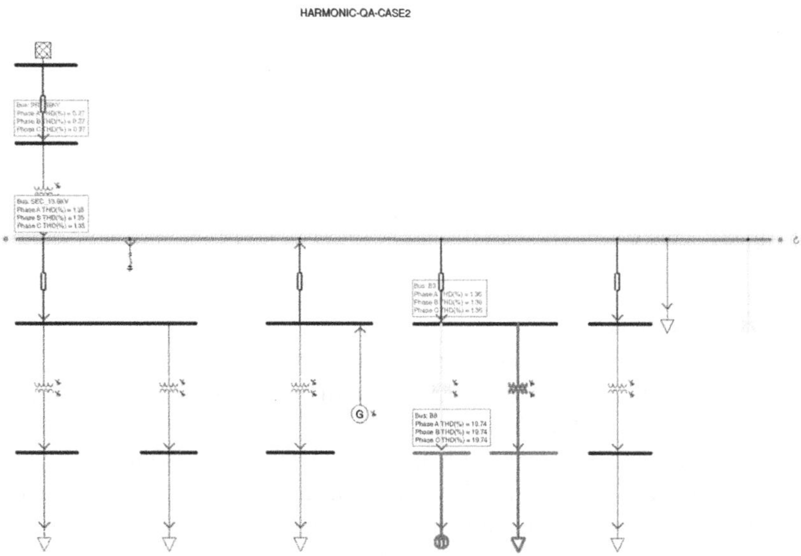

Figure 6.24 Harmonic analysis with the filter results display.

from the previous 5.36% to 1.35%. Plots in Figure 6.25 show the filter effectively reduces fifth as well as seventh harmonics (plot (a)), cleans up and restores sinusoidal waveform on voltages (plot (b)), and removes parallel resonance point around the seventh harmonic (plot (c)). It is noticed that two new parallel resonance points were created due to the system impedance change by adding the filter. However, a close examination of these resonance frequencies finds out they are between 180 and 240 Hz or around 3.5th harmonic, and 540 and 600 Hz or around 9.5th harmonic. Since the harmonic source does not produce these

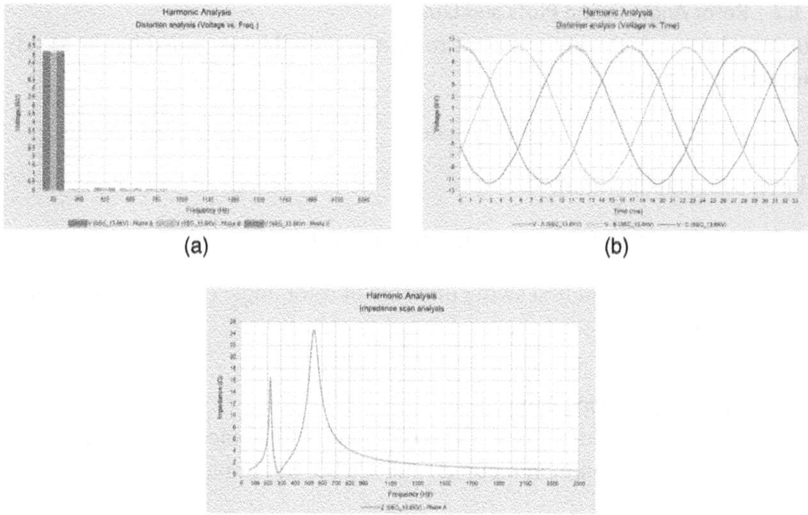

(a) (b)

(c)

Figure 6.25 Plot of harmonic analysis results with the filter: (a) Voltage versus frequency at bus ID SEC_13.8KV. (b) Voltage waveform at bus ID SEC_13.8KV. (c) Impedance frequency scan at bus ID SEC_13.8KV.

harmonics, therefore, the system is safe from resonance and there will not be any voltage magnifying issues.

6.7 Rotor Angle Swing Study

6.7.1 Introduction

6.7.1.1 Causes of Rotor Angle Swings

Many reasons can cause synchronous machine rotor angle swing. But in general, any disturbance that triggers the machine mechanical power on the shaft and electrical power at the terminal (see Swing Equations) out of balance will pull the rotor angle move forward (machine to accelerate in speed) or backward (machine to deaccelerate in speed). If the disturbance is big enough, and the machine does not have sufficient synchronous power as well as damping power to stabilize the rotor angle, rotor angle swing, or oscillation will start and could diverge. Synchronous machine rotor angle swing is a critical measurement for I&C power system stability if they are installed on-site. Transient stability analysis shall be fully performed during the system detailed design phase and rotor angle swings of on-site synchronous machines shall be graphically plotted for careful examination.

6.7.1.2 Rotor Angle Swing Plots and Limits

Both initial rotor angle and transient rotor angle shall be bounded by the steady state stability limit,

$$0 < \delta < 90^{\circ} \qquad \text{at steady state}$$

and transient stability limit,

$$-180^{\circ} < \delta < 180^{\circ} \qquad \text{during transient}$$

Examples of stable and first swing unstable of rotor angle curves are illustrated in Figure 6.26.

6.7.1.3 Rotor Angle Swing Control

The transient stability analysis shall recommend with remedy solutions for the unstable conditions. Common methods include but not limited to:

1) Decide proper critical fault clearing time (CFCT) and critical system separation time (CSST) and set the protections accordingly.
2) Adjust the synchronous generator stability margin.
3) Replace with gas-insulated switchgear for major busbars to prevent faults.
4) Employ and configure out-of-step relays for fast fault clearning.
5) Add differential relays for bus and major equipment for fault protections.
6) Enhance the excitation system for fast reaction.
7) Install and tune power system stabilizer (PSS) for major synchronous generator.
8) Any other protection schemes to detect disturbances, reduce rotor angle transient swing magnitude, and isolate faults quickly.

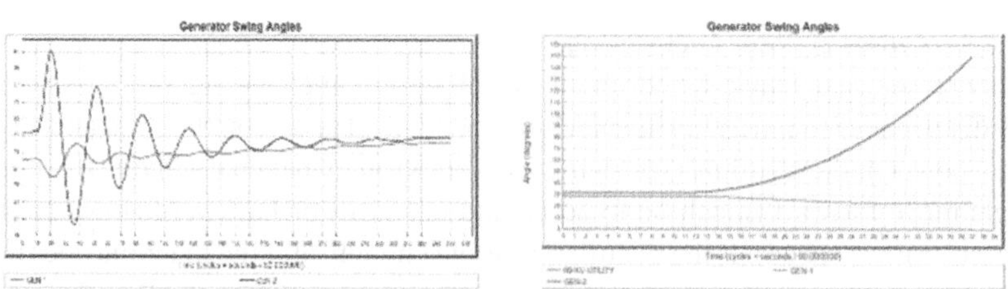

(a) (b)

Figure 6.26 Sample generator rotor angle swing: (a) Stable swing (both). (b) The first swing unstable (blue [refer online]). Source: Dai [3]/with permission of IEEE.

6.7.2 A Rotor Angle Swing Analysis Case

6.7.2.1 System Description

This case study illustrates transient stability analysis in an industrial plant subjecting to a fault [4]. Synchronous generator rotor angle swing under a fault condition is observed and studied. The system single-line diagram shown in Figure 6.27 includes two 25 km 69 kV double circuit transmission lines feeding into a substation and two outgoing 3 km transmission lines providing power to the industrial plant via two-step down 69/13.8 kV, 15 MVA transformers. Two main 13.8 kV generator switchgears distribute power to loads and motors at varying voltages of 4.16, 2.4, and 0.48 kV. The generations of two on-site generators are scheduled at Gen 1 = 11.25 MW, Gen 2 = 10 MW while importing 17.33 MW from the utility.

6.7.2.2 Dynamic Components Model

Dynamic models of the synchronous generator in Figure 6.28, prime mover and speed governor in Figure 6.29, and excitation system in Figure 6.30 are specified using the manufacturer data or as per IEEE recommendations. The data include generator impedances and time constants, prime mover power rating and droop, exciter gains, and all the time constants required for the complete modeling of the control system.

As a common practice, all running induction motors above 50 hp are dynamically modeled in Figure 6.31 as they represent the higher percentage of loads in a typical industrial facility, and motor inertia does contribute to the system response and stability behavior.

Static loads in the system are also modeled dynamically to represent any load variations as a function of voltage and frequency shown Equations (6.1) and (6.2):

$$P(t) = P_0 \times (V_{pu}(t))^{np} \times [1 + P_{freq}(F_{pu}(t) - 1)] \tag{6.1}$$

$$Q(t) = Q_0 \times (V_{pu}(t))^{np} \times [1 + Q_{freq}(F_{pu}(t) - 1)] \tag{6.2}$$

where,

$P(t), Q(t)$ = Active and reactive power at time (t)
$V_{pu}(t)$ = Per unit load bus voltage at time (t)
$F_{pu}(t)$ = Per unit system frequency at time (t)
P_0, Q_0 = Initial active and reactive power of the load
np, nq = Active and reactive power voltage exponent
P_{freq}, Q_{freq} = Active and reactive frequency factors

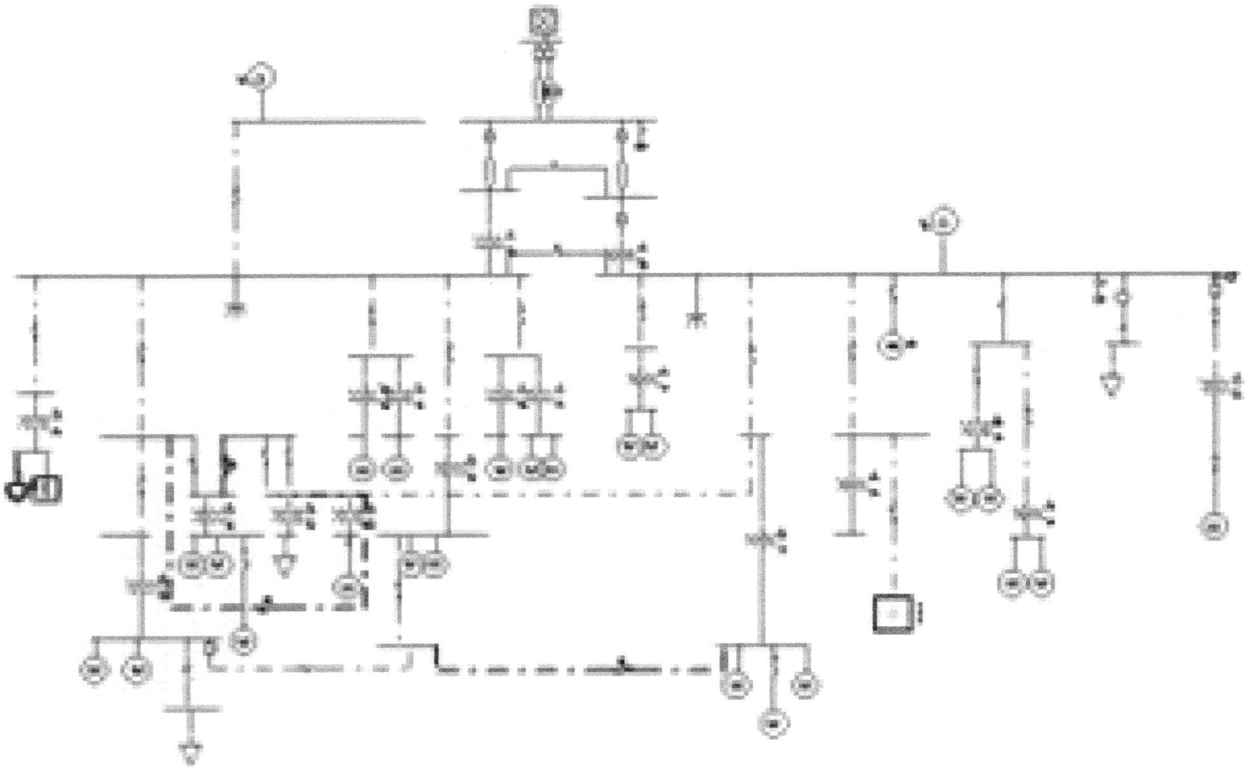

Figure 6.27 Plant single-line diagram. Source: Dai [4]/with permission of IEEE.

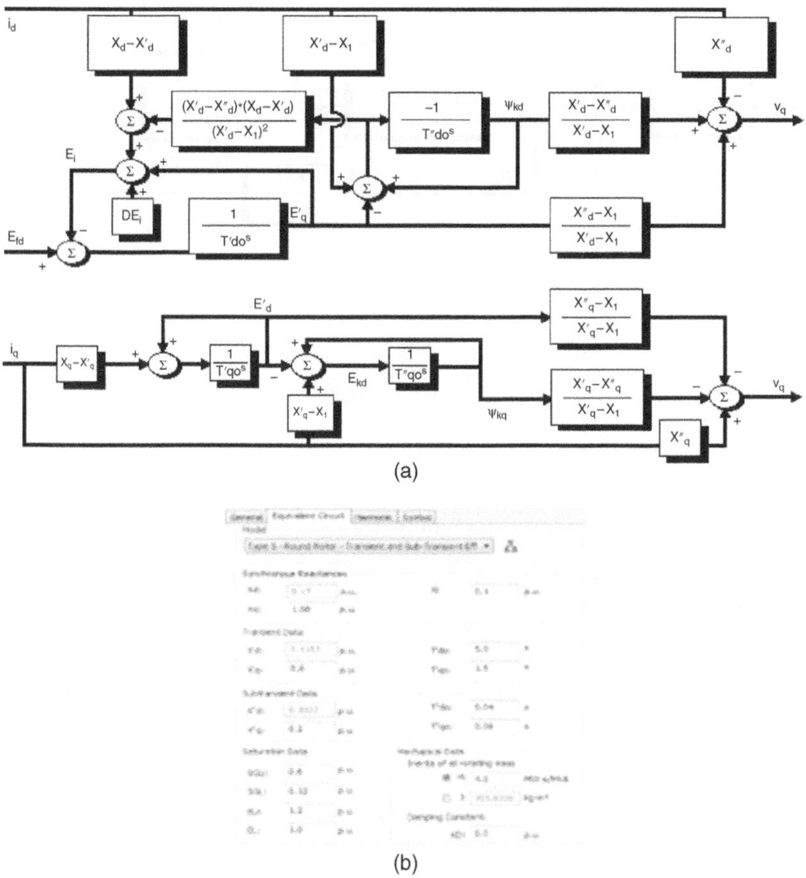

Figure 6.28 Typical synchronous generator model and data. Source: Dai [4]/with permission of IEEE.

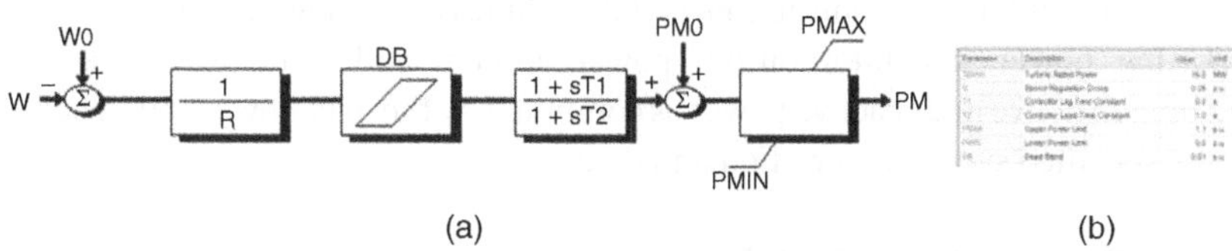

Figure 6.29 Typical speed governor system model and data. Source: Dai [4]/with permission of IEEE.

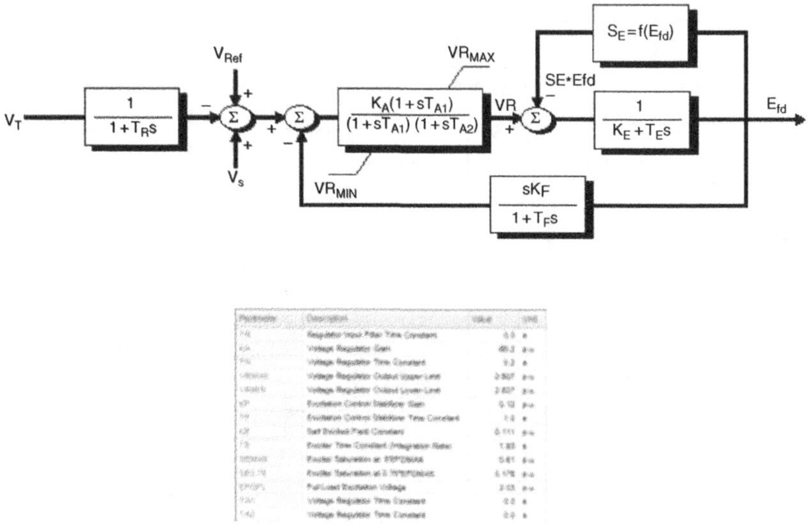

(b)

Figure 6.30 Typical generator excitation system model and data. Source: Dai [4]/with permission of IEEE.

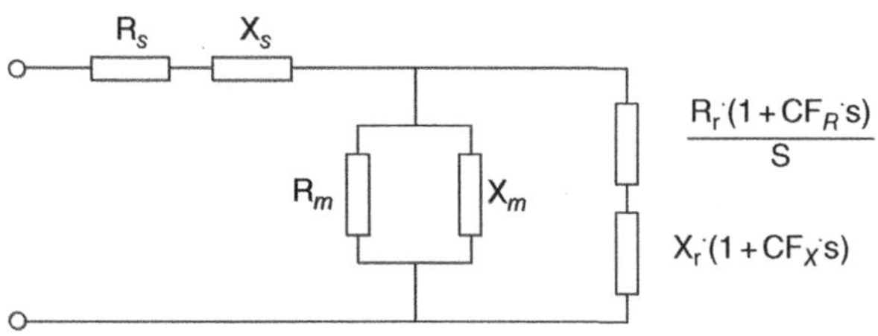

Figure 6.31 Induction motor equivalent circuit model. Source: Dai [4]/with permission of IEEE.

6.7.2.3 Initial Condition

The system's initial condition at time $t = 0.0$ includes bus voltages, grid importing/exporting power, generator output power, motors and static loads loading, and branch active and reactive power as computed by the load flow solution based on the specified system description and data.

6.7.2.4 Transient Stability Study Parameters

The integration time step of the transient simulation loop shall be set to less than the numerical value of the smallest time constant entered in the control parameters of the system. Typically, for I&C power systems, it is in the order of 0.2 cycles. The maximum angular separation beyond which loss of synchronism between generators is a critical parameter. Typically, it is in the range of 120–140°. Finally, the total simulation time is in the order of a few seconds to a tenth of seconds as a typical time span for a transient stability study.

6.7.2.5 Disturbances

Transient stability simulation and analysis typically include following disturbances but are not limited to:

1) Tripping overhead lines and cables with or without automatic reclosure
2) Single pole switching
3) Tripping of synchronous machines
4) Adding and removing shunts
5) Load shedding and load addition
6) Applying and clearing faults with/without a fault impedance (fault types including three-phase, line-to-line, ling-to-line-to-ground, and ling-to-ground)
7) Starting and stopping of single or multiple motors
8) Automatic switching by protective relays

In this study case, the analysis scenario simulates a three-phase fault 15 km away from the plant on the 69 kV transmission lines.

6.7.2.6 Variable Monitoring and Alert

During the simulation, equipment responses and control variables to evaluate the system stability to disturbances and the results are usually displayed in graphical output charts versus time. These responses and variables include generator rotor swing angle, which is a measure of system stability, active and reactive power flows, bus voltages, and system frequency. Based on the monitoring of these responses and variables, the simulation software can provide system-wide alerts for generator rotor angle swings and generator or load bus voltage and frequency variations that exceed predefined thresholds and durations.

6.7.2.7 Simulation and the First Remedy Action

Simulation of the transmission line fault finds the synchronous generator rotor angle lost stability and, consequently, a CFCT was searched. Based on the plot of two generator rotor angle swing curves and measurement of the relative rotor angle between the two in Figure 6.32, it was observed that the CFCT before the loss of synchronism of the plant generators was 10 cycles from the inception of the fault. The first remedy action was then taken at this CFCT point to perform a system separation from the utility.

It was noticed that after the grid separation the frequency of the system continuous to drop below the critical threshold of ±0.5 Hz and did not recover (gray curve in Figure 6.33). This proves that after the fault is isolated from the system,

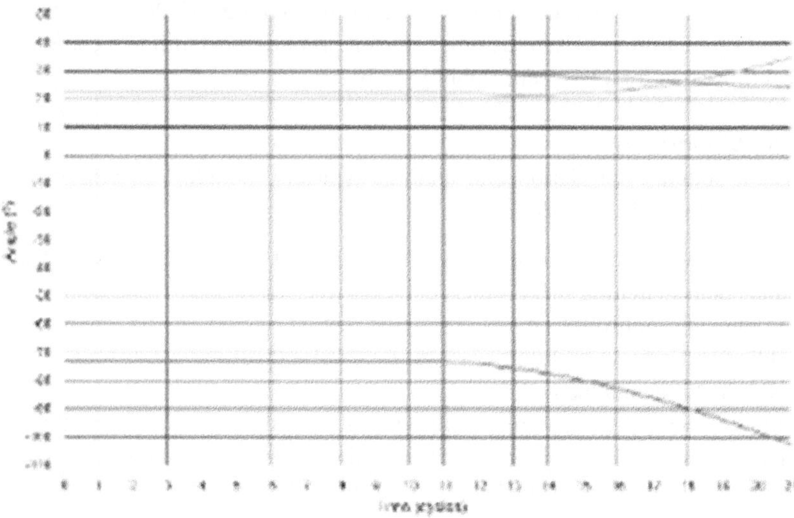

Figure 6.32 Synchronous machine rotor angles swing during fault. Source: Dai [4]/with permission of IEEE.

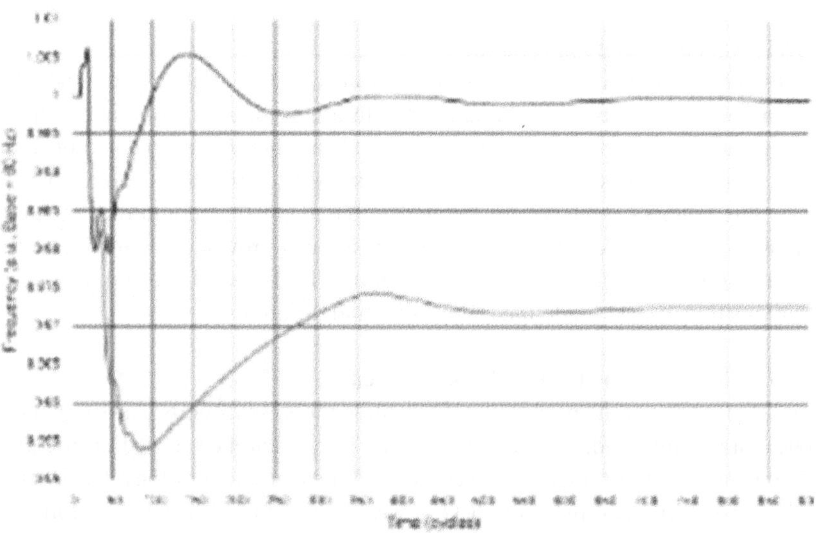

Figure 6.33 Frequency response without load shedding (red line [refer online]) and with load shedding (blue line [refer online]). Source: Dai [4]/with permission of IEEE.

a second remedy action is needed to fully recover the system operational requirement including the frequency stability.

6.7.2.8 The Second Remedy Action and Validation

Now set the frequency relay on the main 13.8 kV bus ID 04:MILL-2 to the following scheme:

1) The under-frequency set point at 0.5 Hz below the system frequency with a total activation time, including breaker speed 4 cycles.
2) An observation delay of six cycles to monitor the system frequency recovery before initiating any selective load shedding.
3) Tripping both breakers CB-7 and CB-8 that connect to 10 MW, 4 Mvar load, and an 8 MVA synchronous motor, respectively.

Ran transient stability simulation with the second remedy for 900 cycles, and the response of the variables of interest was monitored, as shown in Figures 6.33 and 6.34. The system frequency is recovered, as shown in Figure 6.33. The rotor angle separation (relative rotor angle) of GEN 2 with respect to GEN 1 in Figure 6.34a will stabilize at the new operating conditions of the plant. After the selective load shedding, the power output of both plant synchronous generators stabilizes within their predisturbance power generation schedules (Figure 6.34b).

The majority of loads in industrial power systems are induction motors, and it is important to monitor their dynamic response to determine if they can ride through the disturbance and resume normal plant operations. The current contributions from motors during fault conditions are fairly large, and their combined inertia will have an impact on the system response time. The speed of a large 2500 HP, 2.4 kV induction motor is monitored during the transient (Figure 6.35a) to check its reacceleration condition. In addition, the voltage profile of the main buses in the system is also evaluated for nominal operating conditions within the ±5% acceptable limits (Figure 6.35b).

The analysis validated that after sufficient loads are shed, the system can return to a new steady state operating condition, i.e.:

1) System frequency recovers (black line in Figure 6.33).
2) Generator mechanical power and electrical power rebalanced (Figure 6.34b).
3) Induction motor speed after reacceleration stabilized (Figure 6.35a).
4) System voltage dipped during the fault but quickly recovered after the fault was separated (Figure 6.35b).

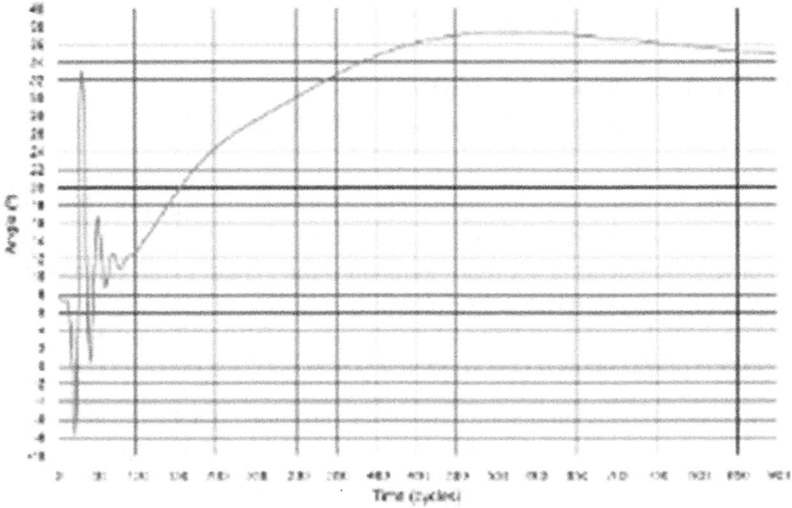

(a)

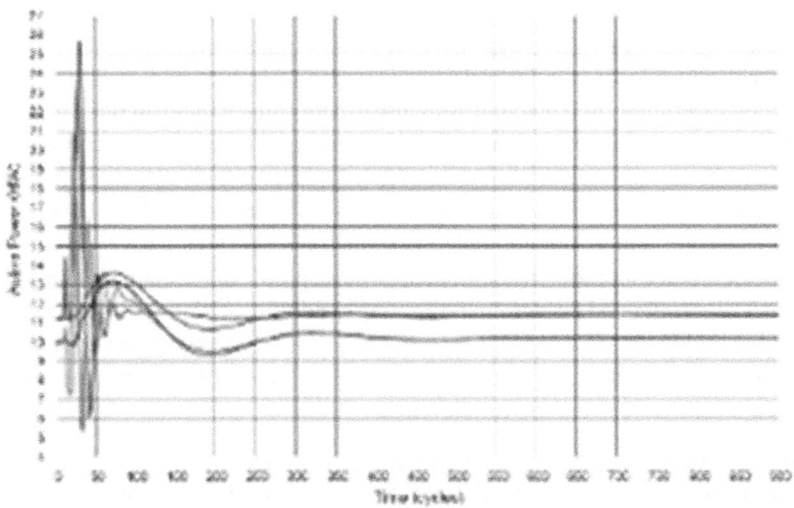

(b)

Figure 6.34 Generator responses with loading shedding: (a) Relative rotor angle swings. (b) Electrical and mechanical power. Source: Dai [4]/with permission of IEEE.

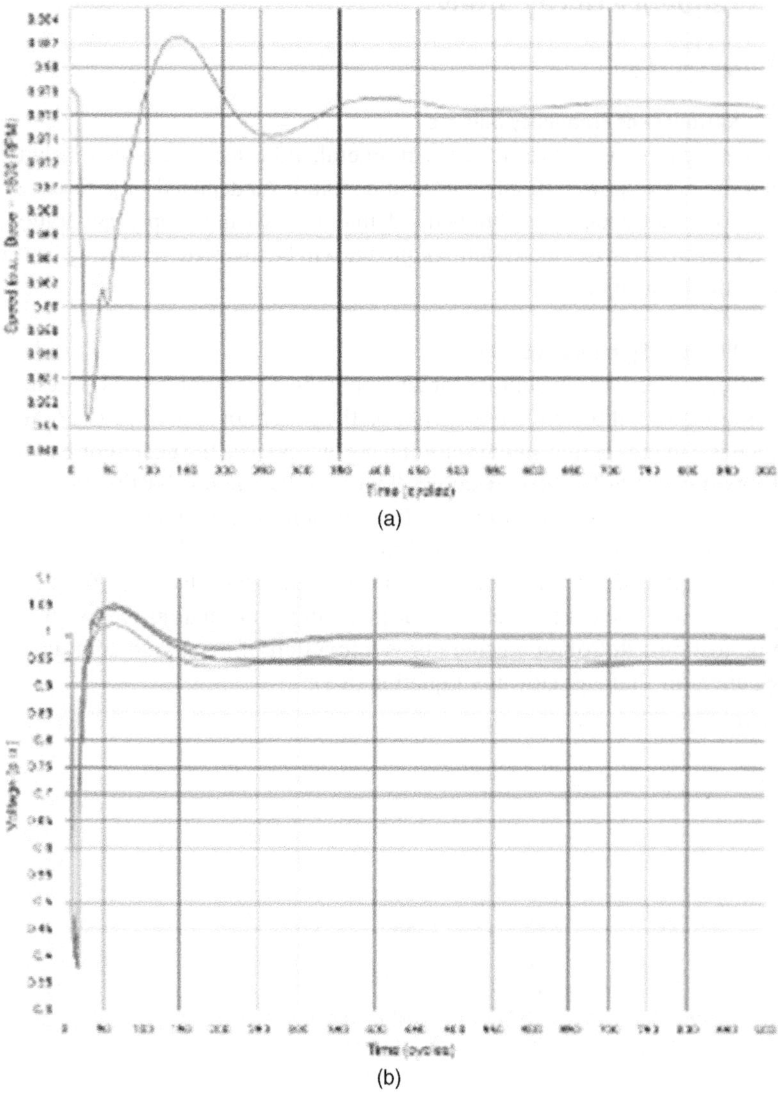

Figure 6.35 Motor speed and bus voltage recovery: (a) The largest induction motor speed. (b) Bus voltage profile. Source: Dai [4]/with permission of IEEE.

6.8 Frequency Decay Study

6.8.1 Introduction

6.8.1.1 Rationale of Frequency Balance

Frequency decay in I&C power systems, in general, indicates there is an imbalance between the active power supply and load demand. As depicted in Figure 6.36, where a nominal frequency is obtained if demand (load) and supply (generation) are in balance; otherwise, the system frequency will be diverted depending on the degree of the imbalance.

6.8.1.2 Effect of Spinning Reserve

A generator has certain capabilities for adjusting the generation output to restore generation-load equilibrium. This depends on the generator's speed-recovery ability, which is a matter of how much spinning power reserved the generator has and speed governor control system dynamics. Figure 6.37a and b show the frequency response curves after an initial dip with different spinning reserves.

In addition to the frequency decay and the lowest value (nadir), rate of change of frequency (RoCoF) is also very important to understand how fast frequency changes after the incident which provides essential information to design and select protections. RoCoF of a generator is a function of net acceleration power on the generator shaft, as shown in Equation (6.3):

$$\frac{df}{dt} = \frac{P_a f_0}{2GH}. \tag{6.3}$$

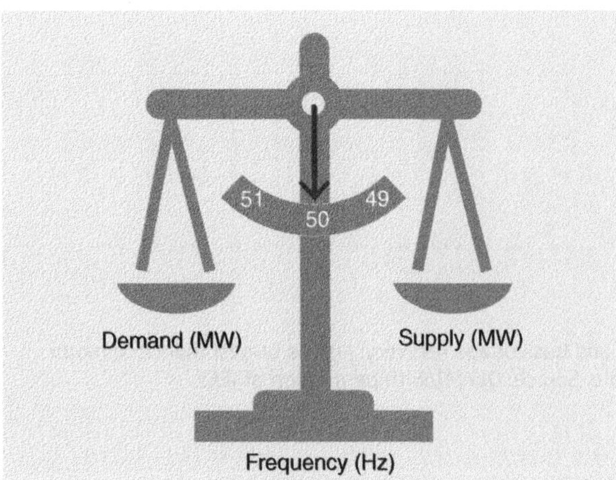

Figure 6.36 Frequency stability is a result of generation and load balance. Source: Dai [5]/with permission of IEEE.

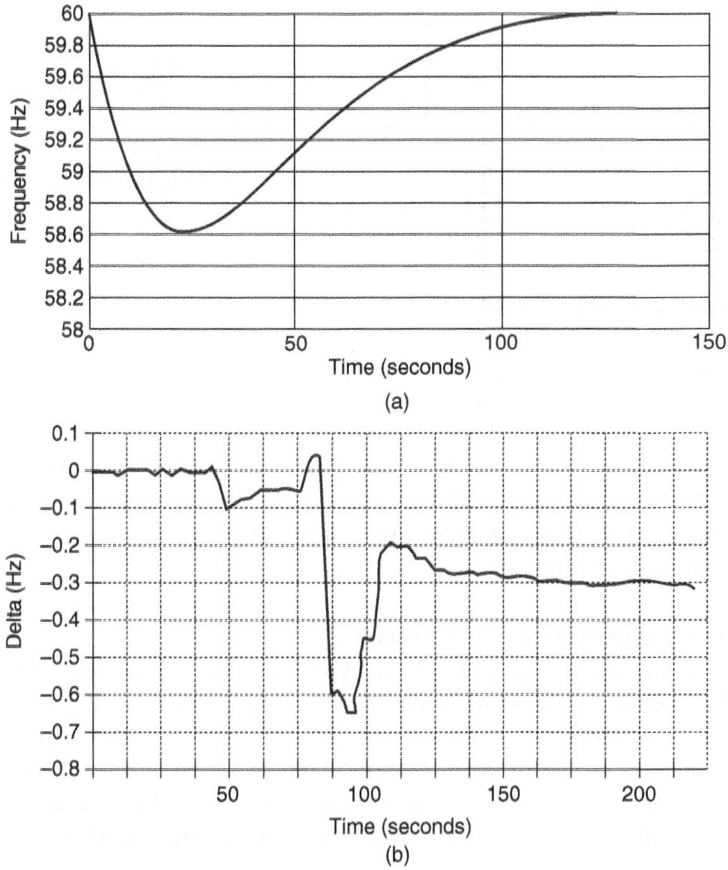

Figure 6.37 Relationship between frequency change and generator spinning reserve: (a) With sufficient spinning reserve. (b) With insufficient spinning reserve. Source: IEEE Standard C37.117 [6]/with permission of IEEE.

where,

P_a = net accelerating power (negative if decelerating).
f_0 = nominal frequency (Hz).
G = nominal MVA of machine.
H = inertia constant (MW-s/MVA).

6.8.2 A Frequency Decay and Protection Case

6.8.2.1 System Description

This case demonstrates a frequency decay case when a local industrial plant's electrical system is disconnected from the utility power grid and becomes islanded.

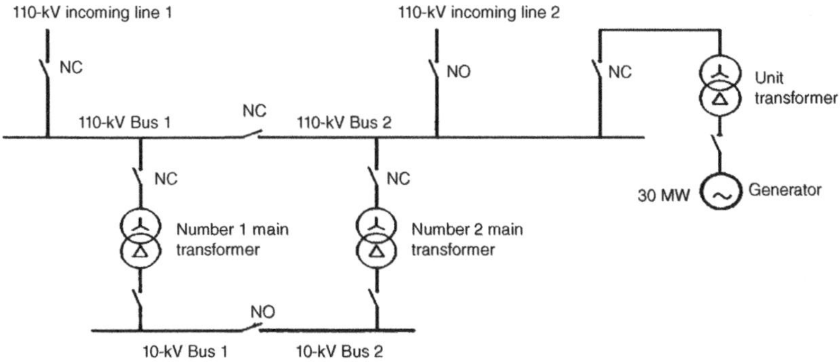

Figure 6.38 Frequency decay analysis case high-level single-line diagram. Source: Dai [5]/with permission of IEEE.

A load-shedding protection is also described. The plant has a normal load of 29 MW, within which around 26 MW are supplied from the in-plant generator, and an additional 3 MW are imported from the utility grid. The system's high-level single-line diagram is shown in Figure 6.38.

6.8.2.2 Study Scenario and Simulation Results

When the islanding condition is initiated, the maximum generator steam turbine mechanical power around 26.5 MW is less than the load demand such that the generator acceleration power becomes negative causing both the generator speed and system frequency start to decay. System dynamic models, including the generator, its excitation, and governor systems, were constructed for frequency decay analysis. Simulation results matched the above expectation and are shown in Figure 6.39.

6.8.2.3 Load Shedding Remedy Development

Since the in-plant generator is not able to support the load and maintain system frequency, a load-shedding scheme needs to be designed. To prepare the load-shedding scheme, load priority must be respected. Table 6.7 lists a general load priority table. Each row is a load group that contains a number of loads defined with one load priority order. Column 1 is the priority order (1 is the lowest), column 2 is the equipment group name, column 3 is the number of load equipment in the group, columns 4 and each higher column (except the last column) list names and loading in kilowatts (kW) for all the loading equipment in that equipment set (organized due to power supply and process dependency), and the last column is the total load expressed in kW for the load group.

The next step is to understand system frequency decay characteristics. A series of simulations were performed at three different initial generation conditions,

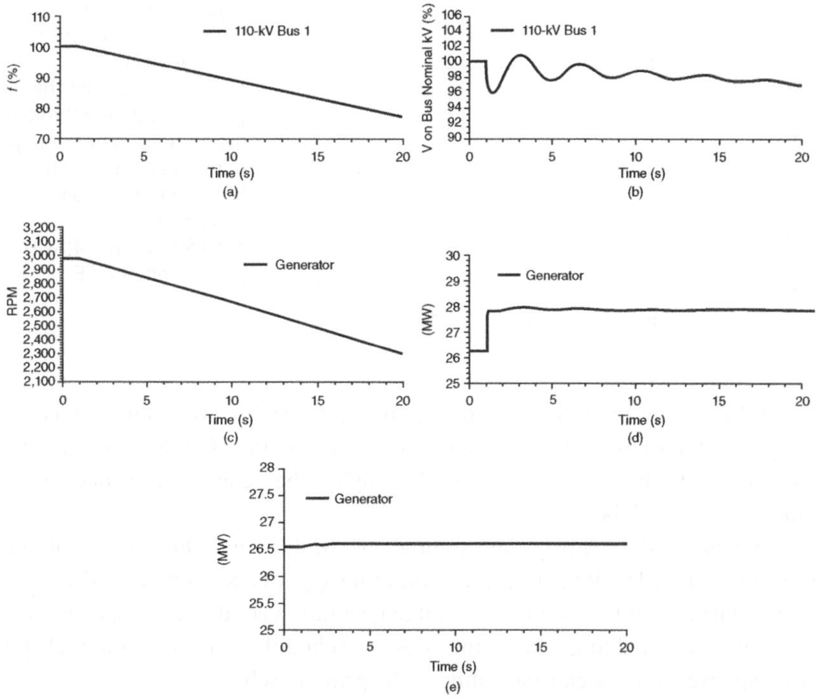

Figure 6.39 The system responds to a sudden loss of gird power import. (a) 110-kV bus frequency (%). (b) 110-kV bus voltage (%). (c) Generator speed (r/min). (d) Generator active output power (MW). and (e) Generator mechanical power (MW). Source: Dai [5]/with permission of IEEE.

Table 6.7 Load priority table.

Order	Equipment name	Equipment number	Equipment 1		Equipment 2		...	Total (kW)
			Name	(kW)	Name	(kW)		
1	Load Group 1	x	xxxx	xxx	xxxx	xxx	...	xxx
2	Load Group 2	x	xxxx	xxx	xxxx	xxx	...	xxx
3	Load Group 3	x	xxxx	xxx	xxxx	xxx	...	xxx
4	Load Group 4	x	xxxx	xxx	xxxx	xxx	...	xxx
5	Load Group 5	x	xxxx	xxx	xxxx	xxx	...	xxx
6	Load Group 6	x	xxxx	xxx	xxxx	xxx	...	xxx
...	...	...	...	...	...	...	...	...

Source: Dai [5]/with permission of IEEE.

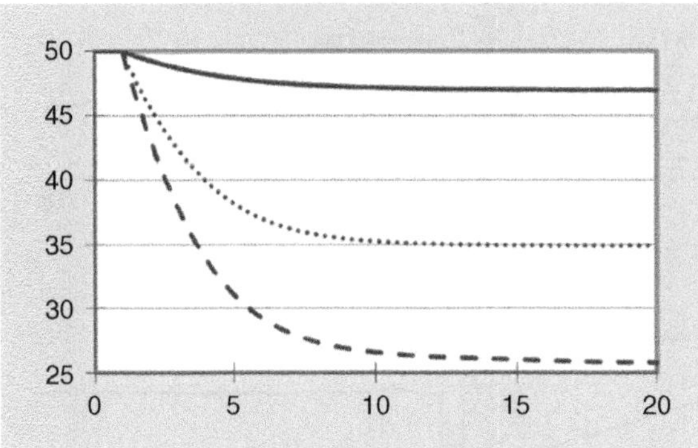

Figure 6.40 The frequency (Hz) versus time (s) curves after disconnecting from the grid at the different initial generations in percentage of the rating (solid line: 87%, dotted line: 64%, and dashed line: 45%).Source: Dai 4]/with permission of IEEE.

and the frequency decay curves corresponding to each initial generations are displayed in Figure 6.40. The frequency decay that occurs after system islanding closely relates to the initial generation. The higher the deficiency, the deeper the frequency decay will be.

The RoCoF df/dt also plays an essential role in deriving a fast load-shedding mitigation action. The RoCoF (Hz/s) versus time (s) curves in Figure 6.41a–c can be derived from Figure 6.40 for three initial generation conditions, respectively.

The final load-shedding scheme includes two schedules, a primary schedule for a fast response, and a backup schedule to the primary schedule.

The primary load-shedding schedule is the first layer protection which is based on frequency and RoCoF and is defined in Table 6.8. This schedule provides the computed load-shedding amount and maximum execution time (Elapsed Time).

Table 6.8 Primary load shedding schedule.

Number	Frequency (Hz)	Rate of change of frequency df/dt (Hz/s)	Elapsed time (s)	Load shed (MW)
1	49	−7.6	0.14	14
2	49	−4.4	0.2	9
3	49	−5.4	1.42	2
4	48.5	−7.43	0.2	14
5	48.5	−4.23	0.32	9
6	48.5	−4.2	2.38	2
7	48	−7.4	0.26	14
8	48	−4.2	0.44	9
9	48	−3.8	3.63	2

Source: Dai [5]/with permission of IEEE.

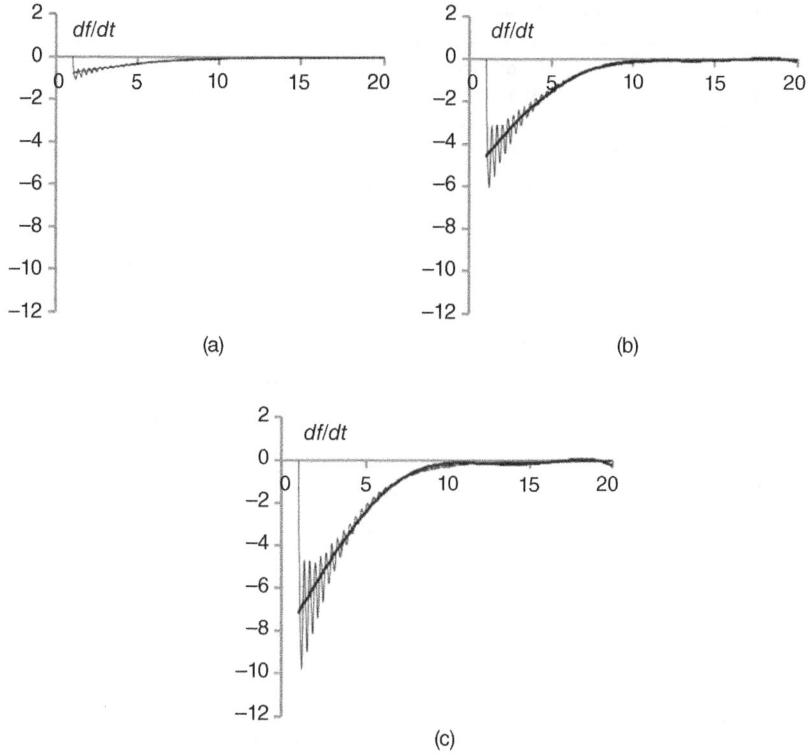

Figure 6.41 The rate of change of frequency (Hz/s) versus time (s) with different initial generation: (a) 87% rating. (b) 64% rating. (c) 45% rating. Source: Dai [5]/with permission of IEEE.

The backup schedule is the second layer of protection that follows rules in Table 6.9. developed based on simulation results in Figures 6.39 and 6.40. This schedule skips any computation and based on this lookup table directly sheds off the amount of load equivalent to the grid importing power prior to the islanding condition to balance the generation and load in an immediate effort.

6.8.2.4 Load Shedding Simulation Validation

The overall proposed load-shedding protection control scheme is replicated and validated through computer simulations using the dynamic system models for simulation. Validation results for the base operation condition are shown in Figure 6.42. It can be seen that after the islanding at one second, load shedding is initiated at approximately three seconds, followed by a full recovery of system frequency and voltage.

Table 6.9 Backup load shedding schedule.

Import power	Frequency (Hz)	Shed time (s)	Shed load
Grid importing power ≤2 MW	49.0	≥1.42	Grid importing power
	48.5	≥2.38	Grid importing power
	48.0	≥3.63	Grid importing power
Grid importing power >2 and ≤ 9 MW	49.0	≥0.20 and < 1.42	Grid importing power
	48.5	≥0.32 and < 2.38	Grid importing power
	48.0	≥0.44 and < 3.63	Grid importing power
Grid importing power >9 and ≤ 14 MW	49.0	≥0.14 and < 0.20	Grid importing power
	48.5	≥0.20 and < 0.32	Grid importing power
	48.0	≥0.26 and < 0.44	Grid importing power
Grid importing power >14 MW	49.0	<0.14	Grid importing power
	48.5	<0.20	Grid importing power
	48.0	<0.26	Grid importing power

Source: Dai [5]/with permission of IEEE.

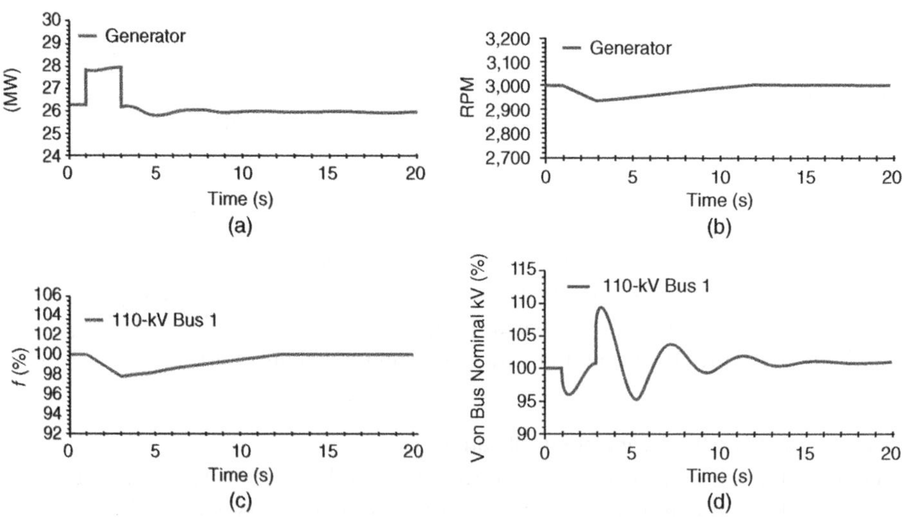

Figure 6.42 The load shedding scheme validation results. (a) Generator active output power (MW). (b) Generator speed (r/min). (c) 110-kV bus 1 frequency (%). (d) 110-kV bus 1 voltage (%). Source: Dai [5]/with permission of IEEE.

References

1 IEEE Std 3002.8™-2018, IEEE Recommended Practice for Conducting Harmonic Studies and Analysis of Industrial and Commercial Power Systems, IEEE, 2018.
2 IEEE Std 519-2022, IEEE Standard for Harmonic Control in Electric Power Systems, IEEE, 2022.
3 JJ Dai, "Practical Approaches to Perform Transient Stability Studies for Industrial and Commercial Power Systems," 2024 IEEE/IAS 60th Industrial and Commercial Power Systems Technical Conference (2023-ICPSD24–0022), May 2024, Las Vegas, NV, USA.
4 JJ Dai, R. Dourian, "Standard Approach to Perform Power System Stability Studies in Oil and Gas Plants," Presented at IEEE PCIC Europe, May 2019, Paris (https://www.scribd.com/document/421660649/Standard-Approach-to-Perform-Power-System-Stability-Studies-in-Oil-and-Gas-Plants?irclickid=zqxWI9yDsxyKU2vw-eSkf2%3A9UkCRGF11UQRTzE0&irpid=2003851&utm_source=impact&utm_medium=cpc&utm_campaign=Scribd_affiliate_p and https://www.semanticscholar.org/paper/STANDARD-APPROACH-TO-PERFORM-POWER-SYSTEM-STABILITY-Dourian/7e95fb0bf65e4f792038ac70c8fee371c6d8d3d1#extracted).
5 JJ, Dai (Sept.-Oct. 2020). Automatic Load-Shedding Protection at a Coal-Chemical Plant. *IEEE Industry Applications Magazine* 26 (5): 77–88.
6 IEEE Standard C37.117, Guide for the Application of Protective Relays Used for Abnormal Frequency Load Shedding and Restoration, IEEE, 2007.

Appendix A

Relevant IEEE Standards and References

The legendary IEEE Std 399-1997 IEEE Recommended Practice for Industrial and Commercial Power Systems Analysis [1], also known as the "Brown Book," was published more than two decades ago. Significant updates and additions are made to the new and upcoming IEEE 3002 standard series in the area of I&C power system analysis. The enhancements in the IEEE Std. 3002 standard series provide more comprehensive guidance on computer modeling and simulation, data and model validation, computer software features and applications, as well as illustrations of system scale and complexity which require enhanced skills and knowledge to perform study and analysis for I&C power systems. By incorporating these standards, this chapter provides an overview of the published IEEE 3002 standard series to ensure that readers are equipped with the latest methodologies and knowledge needed to effectively study and analyze today's sophisticated power systems. This alignment with current industry practices is vital for achieving accuracy, reliability, and efficiency in power system analysis.

A.1 IEEE Std. 3002.2

IEEE Std. 3002.2, titled "IEEE Recommended Practice for Conducting Load-Flow Studies and Analysis of Industrial and Commercial Power Systems," [2] was published in 2018 as its first revision. This standard builds upon Chapter 6, "Load flow studies," from the earlier IEEE Std 399-1997 [1]. After 21 years, significant updates and additions were made to a new standard. The enhancements in IEEE Std. 3002.2 primarily provides more comprehensive guidance on computer modeling and simulation. These updates cover component and network modeling, model validation and verification, simulation data preparation and validation, preparation of computer simulation study cases, and the interpretation of study results.

Industrial and Commercial Power System Analysis Fundamentals and Practice, First Edition. J. J. Dai.

A.1.1 Scope of the Standard

The scope of IEEE Std 3002.2 is to recommend best practices for conducting load-flow studies and analyses for industrial and commercial (I&C) power systems. It encompasses the following main topics:

1) Load flow analysis objectives
2) System modeling for load-flow analysis
3) Required system input data for load-flow analysis
4) Load-flow analysis model and data validation
5) Load-flow simulation methodology
6) Features of load-flow simulation tools
7) Applicable standards for load-flow analysis
8) Load-flow analysis examples
9) Load-flow analysis simulation reports
10) Advanced topics such as optimal power flow, time-domain load flow, and predictive load flow analysis

This standard aims to provide valuable information and guidance to power-oriented engineers with limited experience in load-flow analysis. It serves as a useful resource for engineers involved in the design, study, and operation of I&C power systems.

A.1.2 Load-Flow Analysis Objectives

Load flow is the most fundamental process in power system analysis, essential to every phase of the system life cycle from planning to design and operation. It examines the system's steady-state performance, providing calculated network voltage profiles, branch power and branch current flows, source active and reactive power generation, and load demands under various operating conditions. These conditions include network topology configuration, equipment status, loading levels, and other determining factors.

The standard lists the following common objectives for load-flow analysis:

1) Calculate circuit equipment loadings.
2) Calculate steady-state bus voltages.
3) Calculate network active and reactive power flows.
4) Determine transformer offline and online tap settings.
5) Calculate system active and reactive power losses.
6) Calculate voltage drops across circuit equipment.
7) Calculate active and reactive power exchange with utility sources.
8) Determine generator automatic voltage regulator set points.

9) Identify undervoltage and overvoltage conditions for buses and equipment terminals.
10) Evaluate system performance under planned load conditions, including maximum, normal, minimum, and startup conditions.
11) Evaluate system performance under various planned and unplanned system configurations, such as co-generation on or off, tie-breakers open or closed, utility source connected or disconnected, major distribution equipment in service or out of service, etc.
12) Assess system performance under-credited postcontingencies.
13) Determine power factor correction solutions and settings.

Load-flow analysis is crucial because it provides the following important information to assist in:

1) Verification of system operation under various load and generation conditions
2) Master planning for future growth in loads and generation
3) Configuration of the optimal economic operation for existing systems
4) Establishment of initial conditions for short-circuit, motor starting, harmonic, and transient stability analyses
5) Identification of the need for reactive power support to meet system voltage regulation requirements

By analyzing the solutions under various conditions, load-flow analysis ensures that the power system is designed and operated to satisfy specified performance criteria while improving efficiency, reliability, and cost-effectiveness.

A.1.3 Load-Flow Analysis Methodologies

The standard describes the load-flow model in Y-Bus format and outlines the numerical solution formulation and procedures using the popular Gauss–Seidel and Newton–Raphson iterative techniques. These two techniques are compared in terms of their strengths and weaknesses, particularly focusing on convergence speed and dependence on system initial conditions.

The standard includes modeling of synchronous machine power generation and voltage control, which are key factors in determining the final solution of load-flow calculations. A significant addition to the new IEEE Std 3002.2 is the emphasis on modeling data preparation and model validation. Example system data for various equipment in I&C power systems used in load-flow analysis are provided in the appendix. The standard suggests using operating data and/or benchmarked load-flow calculation data for validation purposes.

Defining and selecting load-flow analysis study cases or scenarios is another new addition to the standard. It describes commonly considered load-flow analysis scenarios, which include variations in loading profiles, generation schedules, and network configurations for normal, maximum, minimum, and special loading conditions. The standard presents a sample industrial power system with load-flow analysis results, including single-line diagrams showing voltages and power and current flows, tabulations for bus voltages, circuit branch flows and generation and load data, and referred warning and alarm information related to violation of ratings and operational requirements.

Since computer simulation software has become the primary tool for I&C power system analysis, the standard summarizes both the minimum required software features and desired optional capabilities.

A.2 IEEE Std. 3002.3

IEEE Std 3002.3™-2018, titled "Recommended Practice for Conducting Short-Circuit Studies and Analysis of Industrial and Commercial Power Systems" [3], was published in 2018 as its first revision. This standard builds upon Chapter 7, "Short-circuit studies" from the earlier IEEE Std 399-1997 [1]. After 21 years, significant updates and additions were made to a new standard. The enhancements in IEEE Std 3002.3 primarily provide more comprehensive guidance on computer modeling and simulation. These updates cover component and network modeling, model validation and verification, simulation data preparation and validation, preparation of computer simulation study cases, interpretation of simulation results and the latest relevant industrial standards.

A.2.1 Scope of the Standard

The scope of IEEE Std 3002.3 is to recommend best practices for conducting short-circuit studies and analysis for I&C power systems. It encompasses the following main topics:

1) Short-circuit analysis objectives
2) Description of short-circuit currents
3) Methods for calculating short-circuit currents
4) Equipment modeling for short-circuit current calculations
5) Short-circuit current calculation based on ANSI (American National Standards Institute)
6) Short-circuit current calculation based on IEC (International Electrotechnical Commission)
7) Required equipment data for short-circuit calculations

8) Validation of short-circuit calculation models and data
9) Short-circuit calculation scenarios and parameters
10) Presentation of short-circuit calculation results and reports
11) Features of short-circuit calculation tools
12) Examples of short-circuit calculations

This standard aims to provide valuable information and guidance to power-oriented engineers with limited experience in short-circuit analysis. It serves as a useful resource for engineers involved in the design, study, and operation of I&C power systems.

A.2.2 Short-Circuit Analysis Objectives

Short-circuit analysis is critical for power system analysis due to the severe impact of short-circuit currents, which can cause interruptions in power system operations and pose safety risks to equipment and personnel. Faults must be quickly detected and removed from the system. Accurately calculated short-circuit currents provide the necessary fault levels to design and install proper protections in both new and existing systems for fault interruption and tripping. These calculations are also essential for determining equipment sizes and ratings for bus bars, cables, transformers, breakers, and fuses.

The standard outlines the following objectives for short-circuit analysis:

1) Verify the protective device's closing and latching capability under maximum fault currents.
2) Verify the protective device's fault current interrupting capability.
3) Verify the equipment's ability to withstand mechanical forces due to maximum fault currents.
4) Verify the equipment's ability to withstand thermal stress due to maximum fault energies.
5) Determine protective relay settings based on relay locations and the equipment to be protected.
6) Determine fault currents for estimating arc fault incident energy.

A.2.3 Short-Circuit Analysis Methodologies

The standard provides a detailed description of short-circuit dynamics, including fault current waveforms as a function of time and various significant values such as peak or maximum, interrupting, 30-cycle, steady state, and DC components. It presents equivalent source models for the power grid, synchronous generators and motors, and induction motors for short-circuit current calculations. These models incorporate variable internal impedances adjusted by multipliers based on source types, source sizes and instances of time.

For unbalanced fault calculations, the standard describes the sequence component method and sequence network with examples using the per-unit system. Transformer models, due to their practical connections and configurations, are particularly complex and are specifically detailed in the standard. In addition to the ANSI method, the standard also covers the IEC method, including its handling of source models, network models, and low-voltage systems.

A comparison of ANSI and IEC short-circuit calculation standards is provided, highlighting differences in equipment modeling and calculation methods. For example, the IEC method's explicit handling of prefault voltage, the adjustment of equivalent internal impedance in the ANSI method versus the decayed fault current magnitude in the IEC method, the genuine network model in ANSI versus configuration- and source location-dependent short-circuit calculation formulas in IEC, and so on.

The standard emphasizes the importance of model data requirements and lists data for each category of equipment and sources that can be gathered from nameplates, datasheets, field measurements, typical or sample data sources, and simulation software libraries. Some typical data validation approaches are also described.

The standard recommends typical short-circuit analysis scenarios involving system operation conditions, generation, and load levels, as well as network configurations. It includes both ANSI and IEC short-circuit calculation examples with sample scenarios and illustrates calculation results on single-line diagram displays, in tabulations, and as warning and alarm windows on rating violations. Additionally, the standard highlights the essential features recommended for simulation tools.

A.3 IEEE Std. 3002.7

IEEE Std 3002.7™-2018, titled "IEEE Recommended Practice for Conducting Motor-Starting Studies and Analysis of Industrial and Commercial Power Systems" [4], was published in 2018 as its first revision. This standard builds upon Chapter 9, "Motor-starting studies," from the previously published IEEE Std 399–1997 [1]. After 21 years, significant updates and additions were incorporated into this new standard. The enhancements in IEEE Std 3002.7 primarily provide more detailed guidance on computer modeling and simulation. These updates cover component and network modeling, detailed motor and starting device modeling, comprehensive starting methods, model validation and verification, simulation data preparation and validation, selection of computer simulation study cases, and interpretation of simulation results.

A.3.1 Scope of the Standard

The scope of IEEE Std 3002.7 is to recommend best practices for conducting motor-starting studies and analyses for I&C power systems. It encompasses the following main topics:

1) Relevant industry standards
2) Motor-starting analysis objectives
3) Motor-starting analysis methodologies
4) System and component modeling for motor-starting analysis
5) Model simulation for motor-starting analysis
6) Commonly used motor-starting methods
7) Motor-starting model data requirements
8) Motor-starting data collection and validation
9) Motor-starting analysis study scenarios
10) Recommended features for motor-starting analysis tools
11) Motor-starting analysis results
12) Motor-starting examples

This standard aims to provide valuable information and guidance to power-oriented engineers with limited experience in motor-starting analysis. It serves as a useful resource for engineers involved in the design, study, and operation of I&C power systems.

A.3.2 Motor-Starting Analysis Objectives

Motor-starting is an essential process for power system analysis. It is required in system planning, design, and operation. Motor-starting studies are performed to determine the system voltage profiles and starting motor voltage drops, starting current and reactive power inrush, and the times needed to start the specified motor for the planned motor-starting schedule. This may involve individual large motors as well as groups of motors starting simultaneously or sequentially. The standard lists the following common objectives for motor-starting analysis:

1) Assess voltage drops due to motor starting and limit them to acceptable levels.
2) Ensure voltage sag indices (time and magnitude) induced by motor starting do not disrupt operating loads.
3) Check and size motor feeders to handle the starting current adequately.
4) Ensure motors are accelerated within acceptable times.
5) Evaluate motor and starting load speed–torque characteristics to allow the motor to start.
6) Closely evaluate thermal capability characteristics to ensure no damage is caused.

7) Ensure motor starting does not cause nuisance tripping.
8) Determine proper motor-starting methods and specify settings and parameters for starting devices.
9) Size motor protective devices to handle the planned motor-starting action.

A.3.3 Motor-Starting Analysis Methodologies

The standard introduces some motor fundamentals related to motor-starting torque and the acceleration process for various motors, including induction motors, synchronous motors, and DC motors. The voltage dip effect during motor start is described by the standard, and critical system voltages that must be maintained at different system levels are presented for reference.

The standard further describes equivalent motor circuit models used for motor-starting analysis and network solutions using load-flow techniques. Both static and dynamic motor-starting simulations are depicted and compared in terms of modeling differences and simulation results. While both simulations can calculate the initial voltage dips at the moment of starting a motor, dynamic simulation provides the full-time-domain solution and considers the complete motor mechanical characteristics and load curve for calculation. However, the latter requires more input data from various system components.

The standard covers a variety of popular and practical motor-starting methods, including their features, such as direct online start, reduced voltage start (including capacitor, wye/delta connection switching, autotransformer, etc.), and power electronics controllers (such as soft starters and variable frequency drives).

On the model data preparation and validation side, the standard lists required data, including network, source, and load components, and recommends model and data validation practices. For analysis scenarios, the standard provides guidance on considering single-motor start, multiple-motor simultaneous start, and sequential start study cases. The standard also presents simulation results in several formats, including single-line diagram displays, time-domain plots, tabulations of various variables, and potential warning and alarm information for operational violations. Essential features of computer tools for motor-starting analysis are illustrated with examples.

A.4 IEEE Std. 3002.8

IEEE Std. 3002.8, titled "IEEE Recommended Practice for Conducting Harmonic Studies and Analysis of Industrial and Commercial Power Systems" [5], was published in 2018 as its first revision. This standard is based on Chapter 10, "Harmonic analysis studies," from the previously published IEEE Std 399–1997 [1].

After 21 years, major changes and additions have been made to the new standard. The enhancements primarily provide more guidance regarding computer modeling and simulation required for harmonic analysis of harmonic sources, network components, and the entire network. This includes interharmonics and their modeling, simulation data requirements and validation, types of harmonic analysis and associated simulation methods, the latest code compliance for PCC (point of common coupling) harmonic distortion limitations, simulation study case selection, harmonic mitigation options and design, and illustrative examples with simulation results.

A.4.1 Scope of the Standard

The scope of IEEE Std 3002.8 is to recommend practices for conducting harmonic studies and analysis for I&C power systems. It describes the basic concepts of harmonics and discusses the analysis needs, data preparations, problem identification, and corrective measures for the analysis. The standard covers the following main topics:

1) Harmonic analysis objectives
2) Harmonic sources and impacts in I&C power systems
3) Harmonic distribution and harmonic resonance
4) Harmonic analysis related standards
5) Harmonic source modeling
6) Harmonic filter modeling
7) Other component and network modeling for harmonic analysis
8) Required data for harmonic analysis and preparation
9) Harmonic model and data validation
10) Harmonic analysis scenario considerations
11) Harmonic load-flow and frequency scan methodologies
12) Features for harmonic analysis tools
13) Harmonic simulation results
14) Harmonic analysis examples

This standard is targeted to provide rich information and value to power-oriented engineers with limited experience in harmonic analysis. It serves as an aid to engineers involved in the design, study, and operation of I&C power systems.

A.4.2 Harmonic Analysis Objectives

With the growing installations of nonlinear loads in I&C power systems, the effects of harmonics on system safety and operation, and their impact on the utility and neighboring loads, always need to be analyzed. Harmonic analysis is an essential requirement during the design, expansion, or operation phases if the system contains significant nonlinear loads, the system exhibits harmonic-related issues, or the unity requires a point of common coupling harmonic distortions report. The study is applicable to the following conditions:

1) Verify and comply with IEEE Std 519.
2) Resolve existing harmonic-related problems, such as equipment failures or misoperation of protective relays or control devices.
3) Design system expansion plans where significant nonlinear loads or capacitance are added.
4) Design new power systems where load flow, power factor compensation, and harmonic analyses are considered as an integrated study.

Objectives for conducting harmonic-analysis studies of I&C power systems include, but are not limited to, the following:

1) Benchmark existing systems and collect data to calibrate system models for future studies.
2) Identify the location, type, and magnitude of harmonic sources.
3) Study harmonic penetration into the system.
4) Simulate and investigate harmonic impacts on system voltages, currents and other influences to equipment performance.
5) Calculate voltage and current harmonic distortions for individual and total harmonic contents.
6) Check for violations in harmonic voltage and current distortion levels.
7) Calculate other harmonic indices and compare them to standard or code limitations.
8) Identify if the system has parallel or series resonance conditions and the resonance frequencies and locations.
9) Design, tune and test harmonic filters.
10) Evaluate transformer phase shift and analyze its effects on harmonic current cancellation.
11) Evaluate and validate other harmonic mitigation designs and their performance.

Depending on the system's design, expansion, or operation phase, harmonic objectives may have different priorities or emphases. Analyzing harmonic simulation results under various conditions can help ensure that the power system is designed and operated to satisfy power quality criteria while improving efficiency, reliability, and cost.

A.4.3 Harmonic Analysis Methodologies

The standard describes harmonic generation and origins, referred to as harmonic sources, which are mainly due to power electronic switching devices and nonlinear loads. It highlights the effects of harmonics, including heating, overloading, losses, negative motor torque, and issues with metering, control, and communication. In addition, the standard illustrates harmonic resonance that can occur in the system due to interaction between inductive components such as motors, which dominantly exist in I&C power systems, and capacitive components, which are mainly from power factor correction capacitors. Parallel or series harmonic resonance can cause extreme overvoltages or overcurrents, damaging equipment and interrupting operations.

The standard presents detailed modeling for harmonic sources, including characteristic harmonics, non-characteristic harmonics, and interharmonics, various power distribution equipment and loads, and the network for simulation study purposes. Popular types of passive harmonic filter models are discussed with their frequency response curves. Required data for each type of model are listed, and data sources and validation options are also discussed.

Both harmonic load flow and harmonic frequency scan simulation solutions are presented in the standard. Typical harmonic study scenarios recommended by the standard cover loading condition variation, system configuration changes, capacitor operation scheduling, and maximum harmonic generation. Harmonic distortion indices from simulation analysis are described, and harmonic voltage and current distortion limits by grids at the interfacing point are explained.

The standard references a wide range of industry standards related to harmonic issues, including IEEE Std 519-2014™, "IEEE Recommended Practices and Requirements for Harmonic Control in Electric Power Systems," [6] and IEEE Std 18™, "IEEE Standard for Shunt Power Capacitors." [7] Essential features from computer simulation tools for harmonic simulations are recommended.

A couple of harmonic analysis cases are presented to demonstrate harmonic distortion reduction and parallel resonance point shift by tuning passive harmonic filter groups and implementing phase-shifting transformers. The standard presents harmonic analysis results in various formats, including single-line diagram displays, voltage and current waveforms, harmonic distortion indices, harmonic content spectrums, tabulations, and warning and alarm information.

A.5 IEEE Standard on Transient Stability Analysis

IEEE Std. P3002.9, "IEEE Recommended Practice for Conducting Transient Stability Studies and Analysis of Industrial and Commercial Power Systems," is currently under development. This standard will be based on Chapter 8, "Stability studies," from the previously published IEEE Std 399-1997 [1]. Significant advances in this technical area over more than two decades since the release of IEEE 399 have necessitated major enhancements to the existing standard.

This section references a recently published IEEE paper, "Practical Approaches to Perform Transient Stability Studies for Industrial and Commercial Power Systems" [8], which provides extensive coverage on this subject. Compared to IEEE 399, this paper includes additional descriptions of transient stability fundamentals, the importance of synchronization power, which is a key factor in maintaining synchronous machine transient stability, detailed discussions on machine models and machine control models, and the model validation methods. It also covers transient stability study scenario considerations, simulation results and potential findings, transient stability margins, stability enhancements, and some instability mitigations. In addition, the paper provides a rich collection of industry standards and best practices, as well as presents a real-world transient stability study case.

A.5.1 Scope of the Paper

This paper summarizes practical approaches to perform I&C power system transient stability analysis, covering the following topics:

1) Transient stability study scope
2) Transient stability fundamentals
3) Network and component modeling for transient stability studies
4) Control modeling for transient stability studies
5) Model validation and verification
6) Transient stability study scenarios
7) Transient stability study results and findings
8) Possible recommendations and mitigations for transient stability study findings
9) A transient stability study project case

This paper aims to assist new and younger engineers who are just beginning to learn about transient stability analysis and understand required skill sets, and to provide additional references for those expanding their knowledge in this subject.

A.5.2 Transient Stability Analysis Objectives

I&C power system transient stability studies focus on the following areas:

1) Model the entire electrical system, particularly synchronous machines with controls, including rotor and stator, excitation, and speed governor systems.
2) Define, simulate and study various normal and abnormal operating and fault conditions.

3) Analyze machine rotor angle oscillations, system frequency and voltage variations, and other dynamic responses.
4) Identify unstable scenarios.
5) Recommend remedies or solutions to the findings.

The overall objective of the power system stability study is to ensure that the electrical system under study can maintain stable operation following severe disturbances.

A.5.3 Transient Stability Analysis Methodologies

This paper first defines the synchronous machine-related transient stability in I&C power systems as rotor angle stability and its time scales, as per the IEEE working group paper [9]. It further describes two fundamental equations that govern dynamic behaviors and relate a rotating machine's mechanical system to its electrical system: the torque equation and the swing equation. An electrical equivalent circuit representing the machine rotor and stator for facilitating simulation studies is illustrated. Based on this equivalent circuit, the power transfer curve and equal area criteria are derived to define steady-state stability and transient stability limits.

This paper presents a detailed discussion of machine dynamic models used in transient stability studies, along with control system models, including excitation and governor systems. Dynamic models for other I&C power system components, such as grids, motors, and loads, are also covered. Great emphasis is placed on the validation of these dynamic models. A collection of industry practices for model verification and validation, using field measurements, lab tests, or computer simulations, is referenced with an actual study case.

This paper lists the mostly considered study scenarios for transient stability analysis of I&C power systems based on reference [1] and the author's extensive industry consulting experience, which provides valuable assistance to system study engineers. General transient stability analysis and study results, along with potential findings, are also outlined for engineers to focus on. Finally, this paper recommends various transient stability mitigation methods from design, protection, and control perspectives. Among these methods, improving the synchronous generator stability margin and configuring out-of-step relays are discussed in detail.

This paper concludes with a comprehensive system design validation project, including a transient stability study for a major industry plant expansion project involving the addition of multiple new generators and new loads [10]. It provides details on system configuration and equipment, operation philosophy, specific transient stability study objectives, major machine and control system models, and transient stability study scenario selections. Simulation and study results are presented and discussed, findings are illustrated, and recommendations to ensure system stability are shared with the readers.

References

1 IEEE (1998). *Std 399-1997, IEEE Recommended Practice for Industrial and Commercial Power Systems Analysis*. IEEE.

2 IEEE (2018). *Std 3002.2™-2018, IEEE Recommended Practice for Conducting Load-Flow Studies and Analysis of Industrial and Commercial Power Systems*. IEEE.

3 IEEE (2018). *Std 3002.3™-2018, IEEE Recommended Practice for Conducting Short-Circuit Studies and Analysis of Industrial and Commercial Power Systems*. IEEE.

4 IEEE (2018). *Std 3002.7™-2018, IEEE Recommended Practice for Conducting Motor-Starting Studies and Analysis of Industrial and Commercial Power Systems*. IEEE.

5 IEEE (2018). *Std 3002.8™-2018, IEEE Recommended Practice for Conducting Harmonic Studies and Analysis of Industrial and Commercial Power Systems*. IEEE.

6 IEEE Std 519–2014™, IEEE Recommended Practices and Requirements for Harmonic Control in Electric Power Systems, 2014, IEEE.

7 IEEE Std 18™, IEEE Standard for Shunt Power Capacitors, 2012, IEEE.

8 J.J. Dai, "Practical Approaches to Perform Transient Stability Studies for Industrial and Commercial Power Systems," IEEE/IAS 60th Industrial and Commercial Power Systems Technical Conference (2023-ICPSD24-0022), 2024.

9 N. Hatziargyriou, J. Milanovic, C. Rahmann, V. Ajjarapu, et al, "Definition and Classification of Power System Stability – Revisited & Extended," *IEEE Transactions on Power Systems*, volume: 36, Issue: 4, July 2021, Pages: 3271–3281.

10 JJ Dai, "A Complete Electrical System Study for a Refinery with Major Expansion," *IEEE Transactions on Industry Applications*, volume: 60, Issue: 1, Pages: 1599–1610, Jan.-Feb. 2024.

Index

Industrial and Commercial Power System Analysis Fundamentals and Practice, First Edition. J. J. Dai.

Printed and bound by CPI Group (UK) Ltd, Croydon, CR0 4YY

17/09/2026

14964331-0003